Scorpion Venom

This book provides a comprehensive overview of scorpion biology and the medical implications of their venoms. It presents the taxonomic classification, anatomy, morphology, and natural habitats of scorpions, detailing their reproductive processes. It further explores the chemical nature of scorpion venom, discussing its composition, toxicity, and physiological effects, as well as its varied functions and mechanisms of action on ion channels. The chapter also focuses on scorpionism, presenting comprehensive epidemiological data and clinical insights from across the globe and reviewing the origin, evolution, and intricate composition of scorpion venom, framing its functional complexity and evolutionary significance. The book also covers the preventative measures and current treatment strategies for scorpion envenomation. It also addresses the limitations of existing antivenom therapies and examines innovative approaches, including the use of pharmaceuticals to enhance treatment protocols. The final chapter provides the promising biomedical applications of scorpion venom toxins across various medical fields. It discusses the therapeutic potential of these toxins in treating a range of human diseases, from cancer and cardiovascular diseases to autoimmune disorders and diabetes. This book is intended for researchers, clinicians, and students of toxicology, pharmacology, and arachnology.

Scorpion Venom

Evolution, Composition, Medical Impact, and Therapeutic Potential

Bhabana Das and Ashis Kumar Mukherjee

CRC Press
Taylor & Francis Group
Boca Raton London New York

CRC Press is an imprint of the
Taylor & Francis Group, an **informa** business

Designed cover image: Das B, Saviola AJ and Mukherjee AK (2021) Biochemical and Proteomic Characterization, and Pharmacological Insights of Indian Red Scorpion Venom Toxins. Front. Pharmacol. 12:710680
DOI: 10.3389/fphar.2021.710680

First edition published 2025
by CRC Press
2385 NW Executive Center Drive, Suite 320, Boca Raton FL 33431

and by CRC Press
4 Park Square, Milton Park, Abingdon, Oxon, OX14 4RN

CRC Press is an imprint of Taylor & Francis Group, LLC

© 2025 Bhabana Das and Ashis Kumar Mukherjee

ISBN: 9781032890159 (hbk)
ISBN: 9781032890197 (pbk)
ISBN: 9781003540816 (ebk)

DOI: 10.1201/9781003540816

Typeset in Times New Roman
by Newgen Publishing UK

Contents

Preface

Scorpions have been portrayed in mythology, folklore, commercial branding, and the visual arts. The scorpion is one of the first to adapt entirely to terrestrial habitats. Scorpionism, or the harm that toxic scorpion venom causes to people, is a global public health issue. While only a tiny proportion of scorpion species threaten humans, the often lethal symptoms resulting from a single sting have led to recognizing scorpionism as a serious health concern in some areas. Although some species have been deemed medically significant, additional scorpion species can potentially harm people. A wide variety of biomolecules (toxins) found in scorpion venom can disrupt the victim's post-envenomation physiological activity, causing neurotoxicity, nephrotoxicity, cardiotoxicity, and hemolytic activity in sting victims.

Geographically, scorpionism varies significantly in terms of both incidence and severity. Therefore, it is critical to determine the principal risk factors and the particular concerns of the local community. The effects of scorpion stings are a significant reason for emergency room visits, especially for young patients and those with weakened immune systems. Except for Antarctica, the regions of Australia, Southern Latin America, South and Southeast Asia, Saharan Africa (North), South Africa, the Near and Middle East, and Mexico are the most vulnerable to scorpion envenomation. Moreover, worldwide clinical data regarding the epidemiology of scorpions is typically underreported; therefore, the actual magnitude of the problem seems to be more severe. Hence, scorpion stings require urgent attention from the health authorities of each country.

The book contains a wide range of topics about scorpions, like the evolution, diversity, and geographical distribution of medically significant scorpions; taxonomic classification, morphology, and habitat of scorpions; evolution and function of scorpion venom; epidemiology, clinical manifestation, and treatment of scorpion envenomation; and therapeutic application of scorpion venom toxins.

This book will be required reading for academics, medical students, and anyone interested in learning about scorpion venom, its function, etc. The readers' curiosity concerning dangerous scorpion stings and treatments for scorpion sting sufferers will be satiated by this book.

We, the writers, wholeheartedly thank everyone who assisted in any way with the writing of this book. We also owe the funding organizations for their kind assistance with our venom and toxin studies. We sincerely believe that readers will enjoy reading this book.

Bhabana Das
Ashis Kumar Mukherjee
Authors

About the Authors

Ashis Kumar Mukherjee, currently serving as the Director of the Institute of Advanced Study in Science and Technology (IASST), Guwahati, did an MSc in Biochemistry from Banaras Hindu University in 1992, PhD in 1998 on Indian cobra and Indian Russell's viper venoms from the Department of Biochemistry, Burdwan Medical College, Burdwan University, and DSc in Biotechnology from Calcutta University on snake venom phospholipase A_2 and protease enzymes in 2017. He has over 30 years of research experience characterizing snake and scorpion venom toxins. His current research interest includes proteomic analysis of snake and scorpion venoms, pharmacological re-assessment of medicinal plants used against snakebite and scorpion sting treatment, quality assessment of commercial antivenom, and cardiovascular and anticancer drug discovery from natural resources, including snake and scorpion venoms. He has received several awards and medals for his academic and research achievements; the most notable is the Visitor's Award for Research in Basic and Applied Sciences from the honorable President of India in 2018. He is a fellow of the Royal Society of Biology, UK, West Bengal Academy of Science and Technology, Kolkata, India, and Indian Academy of Science, Bangalore. Mukherjee has guided 15 research scholars for PhD degrees and 40 MSc dissertations and has published more than 155 research papers in peer-reviewed international and national journals, several book chapters, books, and review articles. He is also a DBT, ICMR, and WHO task force member on preventing and treating snakebite envenoming.

Bhabana Das, currently serving as the Assistant Professor at the Department of Zoology in Devicharan Baruah Girls' College, Jorhat, Assam (affiliated with Dibrugarh University), did an MSc in Zoology from Gauhati University in 2018 and PhD in scorpion venom at the Department of Molecular Biology and Biotechnology, Tezpur University in 2023. Her current research interests include proteomic analysis of scorpion venom, quality and safety assessment of commercial antiscorpion antivenom, and the discovery of a potent drug formulation for better treatment of scorpion stings.

1 The Evolution, Diversity, and Geographical Distribution of Medically Significant Scorpions in the World

1.1 THE TITANS OF THE PALEOZOIC ERA ANCESTRAL SCORPIONS

Scorpions are long, slender arachnids with two appendages on the front of their body – a pair of pincers for gripping – and a segmented, curving tail with a poisonous stinger at the tip. They are primarily viviparous, nocturnal, and cannibalistic animals. *Hadogenes troglodytes*, the rock scorpion native to South Africa, is the longest scorpion in the world. Females can grow up to 23 cm (9 in) in length (Rubio 2000). The most diminutive scorpion in the world is the *Caribbean Microtityus fundorai*, which is 12 nm in size (0.5 in). The average lifespan of scorpions is 3–5 years; however, *Urodacus yashenkoi* scorpion, generally found in Central Australia, can live up to 24 years (Shorthouse and Marples 1982). The scorpions range from Canada and Central Europe to the Southern tip of South America (Tierra del Fuego) and Africa. However, they were accidentally introduced into New Zealand and England. Scorpions have also been found at elevations from sea level to 5,000 m (more than 16,000 ft) in Europe and the North and South American mountains. A few species live north of Southern Canada, Southern Germany, and Russia (Sridhara et al. 2016).

Scorpions are among the earliest animals to fully adapt to the terrestrial environment. Due to the scarcity of their early fossils, it has been challenging to provide definitive answers to critical concerns, such as when and how they evolved to live on land. Fossils of scorpions have been discovered in various strata, including coal deposits from the Carboniferous Period, marine Silurian and estuarine Devonian sediments, and amber (Figure 1.1). Although they possessed book lungs like present terrestrial species, there is still a debate about whether the first scorpions were of marine or terrestrial origin (Howard 2019). According to the information, scorpions initially emerged as aquatic creatures sometime during the Silurian Period, about 450 million years ago. Since then, their morphology has not changed significantly (Dunlop and Webster 1999; Selden, Dunlop, and Edgecombe 1998; Dunlop, Scholtz, and Selden 2013).

DOI: 10.1201/9781003540816-1

a

b

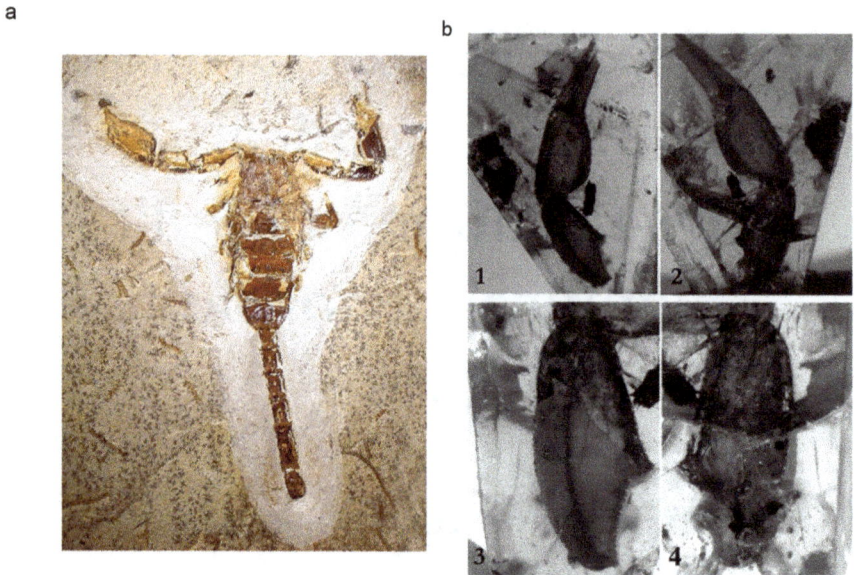

FIGURE 1.1 Scorpion fossils. (a) Early Cretaceous scorpion fossil from Araripe Basin in Brazil. (b) 1–4. Fossil of *Palaeoeuscorpius gallicus* gen. n., sp. n. (1) Pedipalp tibia and patella, dorsal aspect. (2) Idem ventral aspect. (3) Patella, dorsal aspect, showing some trichobothria. (4) Idem, ventral aspect. (a, Reproduced with permission from Lourenço, 2020; CC BY 4.0. b, Figure and legend were adapted with permission from Lourenço, 2003; CC BY – NC – ND.)

There is a significant representation of scorpions in the Paleozoic fossil records. As they share many morphological characteristics, the defunct group known as "sea scorpions" has sparked an ongoing debate on the evolutionary connection between scorpions and Eurypterida throughout history (Kühl et al. 2012; Waddington, Rudkin, and Dunlop 2015; Jeram 1997; Poschmann et al. 2008; Selden and Jeram 1989; Legg et al. 2012). The scorpion *Dolichophonus loudonensis*, which lived in modern-day Scotland during the Silurian, is the oldest fossil discovered in 2021 (Anderson et al. 2021). The Gondwana Scorpio, which lived in the Devonian era, is among the first recorded terrestrial creatures on the Gondwana supercontinent (Gess 2013). In geologic time, the Devonian Period is an interval of the Paleozoic Era that follows the Silurian Period and precedes the Carboniferous Period, spanning about 419.2 million and 358.9 million years ago. Eurypterid-like compound eyes can be seen in Paleozoic scorpions (Schoenemann, Poschmann, and Clarkson 2019). Fossils from the Triassic Period called Protochactas and Protobuthus belong to the current clades Chactoidea and Buthoidea, respectively, demonstrating that the crown group of modern scorpions was already formed by this era (Magnani, Stockar, and Lourenço 2022).

In the early Carboniferous Period (416–359 MYA) or late Devonian Period, the first unquestionably terrestrial (air-breathing) scorpion-like creature was discovered on land (Loret et al. 2001; Lourenço and Gall 2004). These early scorpions, nearly exclusively aquatic or amphibious, eventually diverged into many superfamilies and families. There is a possibility that aquatic (marine) and terrestrial scorpions survived well into the Carboniferous Period (359–299 MYA), and certain species are likely to have survived into the Permian (299–251 MYA) and Triassic (251–200 MYA) eras (Wendruff et al. 2016; Kleffner et al. 2018). All of these extinct scorpions from other planets have been grouped in a single suborder called Branchioscorpionina Kjellesvig-Waering. Along with extant families, fossil scorpions are unambiguously recognized as terrestrial forms; they are grouped into a different suborder called Neoscorpionina by Thorell and Lindström (Sissom 1990; Fet 2000).

According to various authors (Fet 2000; Sissom 1990), 41–47 families and 18–21 superfamilies are in the suborder Branchioscorpionina. The fact that they have so many lineages shows their early and outstanding success. Furthermore, due to the relative scarcity of the fossil record, even 20 superfamilies of scorpions are likely only a tiny part of the total number (Loret et al. 2001; Fet 2000). It seems evident that only a small number of these lineages—possibly just one—survived and gave rise to the current species. Naturally, all currently existing scorpions reside on the land. Strict paleontologists acknowledge many fossil scorpion families, which causes disagreement with neontologists. This difference in approach points to a taxonomic issue, and problems often result from differing methods used in the research conducted by paleontologists and neonatologists (Sissom 1990).

The internal phylogeny of scorpions has long been a source of debate, as has how scorpions are classified among the arachnid orders (Lamoral 1980; Lourenço 1985; Stockwell 1989; Sissom 1990; Soleglad and Fet 2003; Fet and Soleglad 2005; Prendini and Wheeler 2005). Many researchers have traditionally emphasized various physical characteristics of scorpions, leading to different phylogenetic hypotheses; however, none have received widespread acceptance (Fet and Soleglad 2005; Prendini and Wheeler 2005). The family Buthidae and the non-Buthids are physically distinct categories into which scorpions are traditionally separated (Figure 1.2).

The evolutionary position of the relict families Chaerilidae and Pseudochactidae, which overlap features with both groups, has been the subject of particular dispute (Lamoral 1980; Sissom 1990; Coddington et al. 2004; Prendini and Linder 1998). A group that illustrates morphological stability may limit morphological characteristics, which have historically been used to infer higher-level connections between scorpions (and to settle ensuing disagreements) (Prendini and Linder 1998; Prendini and Wheeler 2005). Despite the widespread use of molecular sequence data for phylogenetic reconstruction, no scorpion taxonomy based on molecular information has been proposed (Sharma et al. 2015).

1.2 *PARIOSCORPIO VENATOR* GEN. ET SP. NOV: THE EARLIEST SCORPION THROUGH EVOLUTIONARY ROUTES

Over 2000 identified extant species of scorpions have survived to the present day (Prendini 2001). Among them, the species "*Parioscorpio venator* gen. et sp. Nov" was the earliest exemplified by a tiny exoskeleton showing a unique array of characteristics such as primitive compound eyes and derived (apomorphic) traits including clawed

Tityus bahiensis (Reproduced from Bucaretchi et al., 2014; License no. 5685701455739)

Tityus semulatus (Reproduced from Bucaretchi et al., 2014; License no. 5685701455739)

Androctonus crassicauda (Reproduced from Ghavami et al., 2022; CC BY-NC 4.0)

Parabuthus granulatus (Reproduced from Marks et al., 2019; CC BY-NC-ND 4.0)

Parabuthus transvaalicus (Reproduced from Marks et al., 2019; CC BY-NC-ND 4.0)

Androctonus australis (Reproduced from Sadine et al., 2014; CC BY 4.0)

Centruroides noxius (Reproduced from Teruel et al., 2015; CC BY-NC-ND 4.0)

Mesobuthus tamulus (Reproduced from Das et al., 2021; CC BY 4.0)

Mesobuthus eupeus (Reproduced from Mans, 2017; CC BY-NC 4.0)

Leiurus quinquestriatus (Reproduced from Lowe et al., 2014; CC BY 3.0)

Hottentotta judaicus (Reproduced from Hauke et al., 2017; License no. 5685971190083)

FIGURE 1.2   Scorpions of the Buthidae family.

pedipalps and a small metasoma that ends in a stinger (Figures 1.3 and 1.4). Since the study of scorpion venoms began more than a century ago, some intriguing findings have been uncovered (Casewell et al. 2013; Yang et al. 2017). At the same time, some attempts have been made to elucidate specific potential evolutionary pathways, but the evolutionary significance of mammal-specific toxins is largely unresolved. The Waukesha Lagerstätte is located in the *Pterospathodus eopennatus* superzone according to conodont biostratigraphic zonation, and this biota was believed to be the fossil site for the scorpion species *P. venator* gen. et sp. nov (Kleffner et al. 2018).

This superzone corresponds to the Telychianera during the Silurian period in the geological timescale. The Waukesha Biota, often referred to as the Waukesha Lagerstätte, Brandon Bridge Lagerstätte, or Brandon Bridge fauna, is a vital fossil site that can be found near the Wisconsin cities of Waukesha and Franklin in Milwaukee County. The Brandon Bridge Formation, from the early Silurian era, contains specific layers where this biota is still present (Mikulic, Briggs, and Kluessendorf 1985). It is renowned for exceptionally well-preserved soft-bodied animals, including many species not found in other rocks of comparable age (Jones, Feldmann, and Schweitzer 2015). When the site's discovery was made public in 1985, numerous discoveries followed (Mikulic, Briggs, and Kluessendorf 1985; Jones, Feldmann, and Schweitzer 2015; Wendruff et al. 2020).

Previously, the earliest documented scorpion eurypterid (formally called sea scorpion) was *Dolichophonus loudonensis* from the Beds or Deerhope formation, which consists of laminated fossil siltstones, mudstones, and shales in the Pentland Hills of Scotland (Laurie 1900). The *D. loudonensis* specimens were discovered close

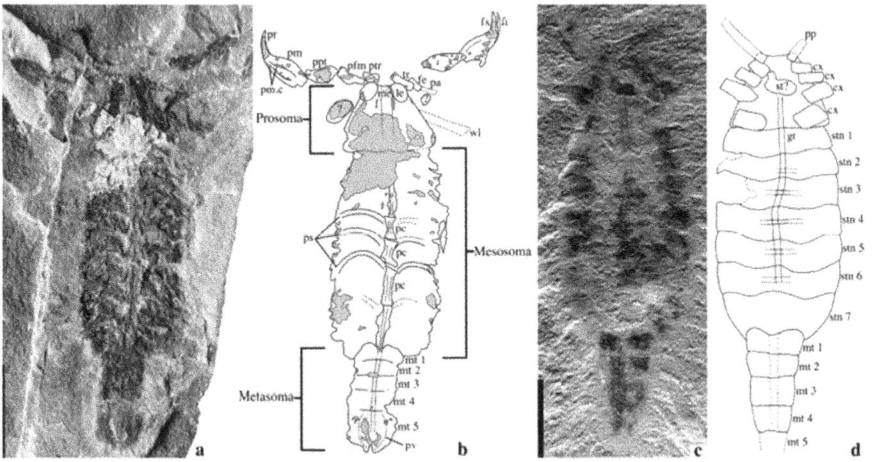

FIGURE 1.3 *Parioscorpio venator* gen. et sp. nov., Brandon Bridge Formation (Silurian), Wisconsin, USA. (a) Holotype, UWGM 2162, photographed under low-angle lighting and revealing internal anatomy; (b) interpretive drawing of holotype; (c) Paratype, UWGM 2163, photographed under low-angle lighting; (d) interpretive drawing of paratype. Abbreviations: cx, coxa; fe, femur; fr, free finger; fx, fixed finger; gt, gut; le, lateral eye; me, median eyes; mt, metasomal segment; pa, patella; pc, pericardium; pfm, pedipalp femur; pm, pedipalpmanus; pm.c, pedipalpmanus carina; ppt, pedipalp patella; pr, pedipalp ramus; ps, pulmo-pericardial sinus; ptr, pedipalp trochanter; pv, poison vesicle; st, sternum; stn, sternite; tr, trochanter; wl, walking leg. The scale bar equals 5 mm. (Figure and legend were adapted with permission from Wendruff et al., 2020; CC BY-NC 4.0.)

to graptolites (a group of colonial animals and their fossils found from the Middle Cambrian through the Lower Carboniferous), indicating the middle *Oktavites spiralis* zone (*Oktavites spiralis* is one of the Telychian Age's index graptolite species (late Llandovery, Silurian) to the middle *Cyrtograptus lapworthi* zone (*C. lapworthi* is also the graptolite species of Telychian Age found near Dammark) and conodonts belonging to *Pterospathodus amorphognathoides* (Bull and Loydell 1995; Loydell 2005).

According to conodont and graptolite data, the *P. amorphognathoides* zone (*P. celonisuperzone*; Telechian stage, c. 435.5–434.5 Ma) lies above the *P. eopennatus* superzone, which corresponds to the strata where *D. loudonensis* is found. *P. venator* gen. et sp. nov. is marked by a tiny exoskeleton presenting a unique character array. Based on earlier research, several characteristics—like compound eyes—are regarded as primitive (plesiomorphic) for arachnids (Selden, Dunlop, and Edgecombe 1998; Loret et al. 2001; Dunlop 2010; Dunlop, Erik Tetlie, and Prendini 2008; Hjelle 1990; Sissom 1990; Petrunkevitch 1953). Other characteristics, such as pedipalps with claws and a thin metasoma, are derived from a stinger (apomorphic). The species *P. venator*, which belongs to the extinct arthropod genus *Parioscorpio*, was discovered in the Silurian-aged Waukesha biota of the Brandon Bridge Formation near Waukesha, Wisconsin, USA. This animal has had a perplexing taxonomic history, passing by many names like spider, crab, and antipodean arthropod (Mikulic,

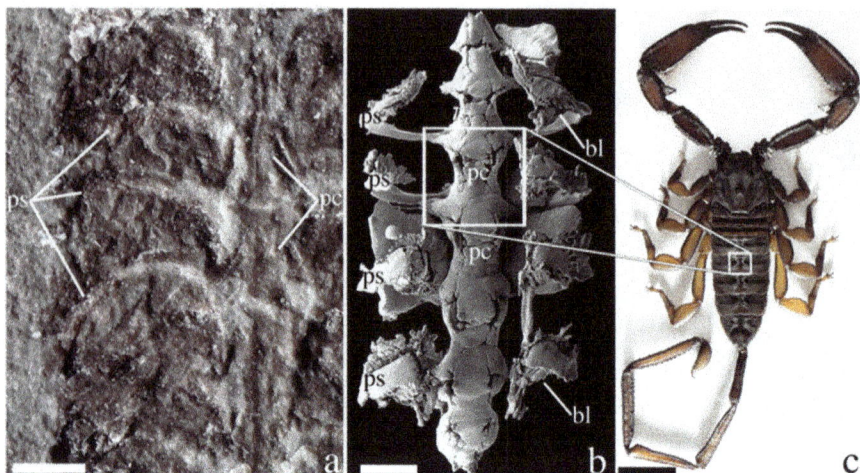

FIGURE 1.4 Medial structures associated with the pulmonary–cardiovascular system in Silurian (a) and Holocene (b, c) scorpions. (a) *Parioscorpio venator* gen. et sp. nov., holotype, detail of medial region showing pulmo-cardiovascular structures; (b) SEM of *Centruroides exilicauda*, corrosion cast of pericardium and associated pulmo-pericardial sinuses. (c) Hadogenes troglodytes, male, dorsal surface, showing medial structures externally reflecting the position of the internal pericardium (compare with b). Abbreviations: bl, book lungs; pc, pericardium; ps, pulmo-pericardial sinus. Scale bars equal 1 mm for (a, b); scale bar equals 1 cm for (c). (Figure and legend were adapted from with permission Wendruff et al. 2020; CC BY-NC 4.0.)

Briggs, and Kluessendorf 1985; Braddy and Dunlop 2021; Wendruff et al. 2020). It is believed that *P. venator* is the only species with a mesosoma with seven tergites and sternites (Figures 1.3 and 1.4). The number of sternites in Paleozoic scorpions has been declining through time.

1.3 THE SISTER SCORPIONS IN EVOLUTIONARY PERIODS

Two younger Silurian species, *Proscorpius osborni* (Dunlop, Erik Tetlie, and Prendini 2008) and *Eramoscorpius brucensis* (Waddington, Rudkin, and Dunlop 2015), had six sternites and lived during the Silurian and Devonian periods (about 422.9–416.0 Ma). Most living and extinct scorpions have four sternites (Hjelle 1990; Sissom 1990), a characteristic developed by at least the Carboniferous Period, the fifth interval of the Paleozoic era (350–200 million years ago). *Allopalaeophonus caledonicus* (Petrunkevitch 1953), *Palaeophonus nuncius*, and *P. osborni* (Dunlop, Erik Tetlie, and Prendini 2008) are younger Silurian species with broad, anterolateral eyes and small medial eyes that are positioned anteromedially. All existing or live scorpions have most Paleozoic forms (e.g., *P. venator*), which lack chemosensory organs and pectins. It is comparable to other, more advanced scorpion taxa in terms of its exterior morphology, as those pectins are rapidly shed in the event of death or molting (McCoy and Brandt 2009; Scholtz and Kamenz 2006).

For scorpions, a telson with an enlarged space for the venom vesicle and sting is present in an asymmetrical position (Dunlop 2010). The partial telson, folded under the fifth metasomal segment, is preserved in the *P. venator* holotype. The more terminal stinger is not visible, but the proximal section exhibits a swelling that is thought to represent a venom vesicle close to the joint with the metasoma. The internal anatomy of *P. venator* is visible in both specimens. The paratype's medial structure, interpreted as the digestive system and resembling that of living scorpions, is visible upon separating the rock through the fossil (Figures 1.3 and 1.4). It is a short, straight tube that connects the prosoma to the metasoma. The central architecture of living scorpions' circulatory and respiratory systems has been extensively studied. It resembles elements conserved in *P. venator* (Figure 1.3a–c). Tepericardium is a network of slender, medial hourglass-shaped structures in the mesosoma that encircles and protects the heart. The pericardium produces strut-like pulmo-pericardial sinuses that extend laterally. These internal medial features and the tergite boundaries are externally reflected on the dorsal cuticle of some living scorpions (Figure 1.4c).

In living scorpions, the lungs and circulatory system are connected by the pulmopericardial canals. Hemolymph, or "blood," is oxygenated by the book lungs and sent to the pericardium. It may be concluded that the organs of respiratory–cardiac architecture were conservative in their evolutionary development. The only extant chelicerates that process oxygen from the air are typically terrestrial species like scorpions (e.g., utilizing book lungs). Marine xiphosurans (horseshoe crabs) can still breathe when they travel to land to reproduce, although they generally draw oxygen from water using external book gills (Botton et al. 1996). Horseshoe crabs (*Xiphosurans*) have respiratory and circulatory systems as complex as scorpions, which could help explain how they can breathe air and survive on land. Arachnid and xiphosuran ancestors probably had a comparable ability to land on the ground. Anatomical characteristics preserved in *P. venator* relate to an early time in the evolutionary history of arachnids when the physiological adjustments required to enable a marine-to-terrestrial transition occurred. It is uncertain if *P. venator* was a fully terrestrial arthropod. The preserved pulmonary–cardiovascular structures' striking resemblance to those of living scorpions and horseshoe crabs suggests that they may have spent considerable time on land (Wendruff et al. 2020).

1.4 GEOGRAPHICAL DIVERSITY AND VARIATION IN VENOM COMPOSITION OF SCORPIONS

The most prominent family of scorpions, Buthidae, is found worldwide, except in New Zealand and Antarctica (Mullen and Sissom 2019). There are 1,225 species in this family (as of January 22, 2021; Rein 2020), of which approximately 50 are deemed harmful to humans (Lourenço 2016). The genera *Androctonus*, *Buthus*, *Centruroides*, *Hottentota*, *Leiurus*, *Mesobuthus*, *Parabuthus*, and *Tityus* comprise the Buthidae C. L. Koch family, which was founded by Carl Ludwig Koch in 1837 and accounts for over 95% of scorpion stings (Lourenço 2018).

Members of the Chactidae family of scorpions can be locally and mildly harmful to humans (Nishikawa et al. 1994; Goyffon and Kovoor 1978). However, more scorpion species can harm humans; some species have been considered medically significant

(Figure 1.2). As of right now, Brazil is home to approximately 160 species, 23 genera, and 4 families of scorpions (Buthidae, Hormuridae, Chactidae, and Buthriuridae), which account for 6.3% of the variety of these arachnids globally (Hauke and Herzig 2017; Pezzi et al. 2019; Ward, Ellsworth, and Nystrom 2018). The majority of the scorpion species found in Brazil belong to the genus *Tityus*. This genus accounts for more than 60% of scorpions found in tropical and subtropical areas (Saúde 2001). The primary sources of scorpionism in people in Brazil are the three *Tityus* species: *Tityus stigmurus* (Brazilian Scorpion), *Tityus bahiensis* (brown scorpion), and *Tityus serrulatus* (yellow scorpion) (Bucaretchi et al. 1995; Eickstedt et al. 1996).

The most severe incidents are caused by the Brazilian scorpion *Tityus serrulatus*, which has a 1% death rate in youngsters and elderly adults. According to Saúde (2001), this species is extensively dispersed over the nation, having been found in the states of Rio de Janeiro, São Paulo, Minas Gerais, Bahia, Espírito Santo, Goiás, and Paraná (Figure 1.5). The ability to reproduce by parthenogenesis (Lourenço 2008) is

FIGURE 1.5 The geographical distribution of *Tityus serrulatus* throughout Brazil. (Image sketched by B. Das.)

one of the elements contributing to its proliferation and dissemination, which makes control over these arachnids more difficult. Another clinically significant scorpion species is *T. stigmurus*, primarily found in the northeastern section of the nation, which is likewise capable of parthenogenesis.

Research carried out in Bahia, Brazil, between 1990 and 1995 revealed that *T. stigmurus* is the primary cause of scorpionism in the area (Lira-da-Silva, Amorim, and Brazil 2000). However, the true epidemiological impact of these scorpionism episodes in these regions is still somewhat unknown due to a lack of information. Notably, *T. bahiensis* shows crossbreeding, which necessitates male and female encounters during particular times of the year, in contrast to the previously described species. The states of Goiás, Mato Grosso do Sul, Minas Gerais, Paraná, Rio Grande do Sul, São Paulo, and Santa Catarina are home to this scorpion (Figure 1.6) (Porto, Brazil, and Lira-da-Silva 2010).

FIGURE 1.6 The geographical distribution of *Tytus bahiensis* throughout Brazil. (Image sketched by B. Das.)

Scorpions have been long known in Iran due to their stings. Owing to climate and weather, Iran is rich in arthropods, especially scorpions (Dehghani and Fathi 2012). Iran is where various types of scorpions and other dangerous species are found. Currently, the primary method of treating scorpion stings (of any species) in the world, including Iran, is to use antiscorpion antivenom, which is questionable in treating some sting cases (Sedaghat and Moghadam 2012).

The impact of the two hazardous groups of scorpions present in Iran, Buthidae, and Hemiscorpiidae, on humans varies due to differences in their venom composition. For instance, species in the Buthidae family primarily cause neurotoxic effects, while Hemiscorpiidae species, including *Hemiscorpius lepturus* (known locally as Gadim), cause hemolytic effects that increase the country's fatality rate (Pipelzadeh et al. 2007; Ghafourian et al. 2016).

Threatened species of scorpions inhabit some areas in different parts of Iran, some of which are fatal to humans (Dehghani and Fathi 2012). More than 42,500 cases of scorpion stings have been recorded per year in Iran, with a mortality rate of 44.7 per 100,000. These figures place the country second only to Mexico in terms of scorpion bite fatality rate (Dehghani and Fathi 2012; Dehghani and Valayi 2005). The actual numbers of scorpion stings and mortality rates are more critical than the yearly recorded figures. Hence, these numbers are not conclusive.

The black scorpion (*Androctonus crassicauda*) in Kashan City was described by Olivier in 1807 during his study of the scorpion fauna in Iran (Olivier 1807). Since then, numerous investigations have identified 53 species, 19 genera, and 4 families of scorpions in Iran. The Buthidae family is the most well-known, with 15 genera and 45 species (Farzanpay 1990; Yousef Mogaddam et al. 2017). *Mesobuthus eupeus, A. crassicauda, H. lepturus, Odontobuthus doriae, Hottentotta saulcyi, Hottentotta schach, Compsobuthus matthiesseni,* and *Mesobuthus caucasicus* are the most medically necessary species, whereas *H. lepturus* and *A. crassicauda* are the most lethal and hazardous to people of Iran (Dehghani and Fathi 2012; Vazirianzadeh and Salahshoor 2015).

According to a study carried out in the northwest of Iran in the provinces of Ardabil, West Azerbaijan, and East Azerbaijan, 368 distinct species of scorpions were found in various habitats and locations with a range of climates in both urban and rural areas. Their preferred habitats include grasslands, abandoned houses, cemeteries, under stones and bark, on farms and plows, and next to homes. Most of these scorpions belong to two families, five genera, and five species: *M. eupeus* (80.16%), *A. crassicauda* (10.60%), *M. (Olivierus) caucasicus* (4.89%), *Scorpio maurus* (2.99%), and *H. (Buthotus) saulcyi* (1.35%). Except for *S. maurus* (Scorpionidae), the other venomous species of scorpions in this region belong to the Buthidae family.

In India, the Indian red scorpion (*Mesobuthus tamulus*) and the Indian black scorpion (*Heterometrus bangalensis*) are most prevalent. Clinical findings that are now available indicate that the venom of Indian red scorpions is more lethal than that of Indian black scorpions (Bawaskar and Bawaskar 1998). Studies also show that immediate medical attention is necessary after a scorpion sting. In India, the most medically essential species are those of the genera *Androctonus, Buthus, Centruroides, Hottentotta, Leiurus, Mesobuthus, Parabuthus,* and *Tityus* (Bawaskar and Bawaskar

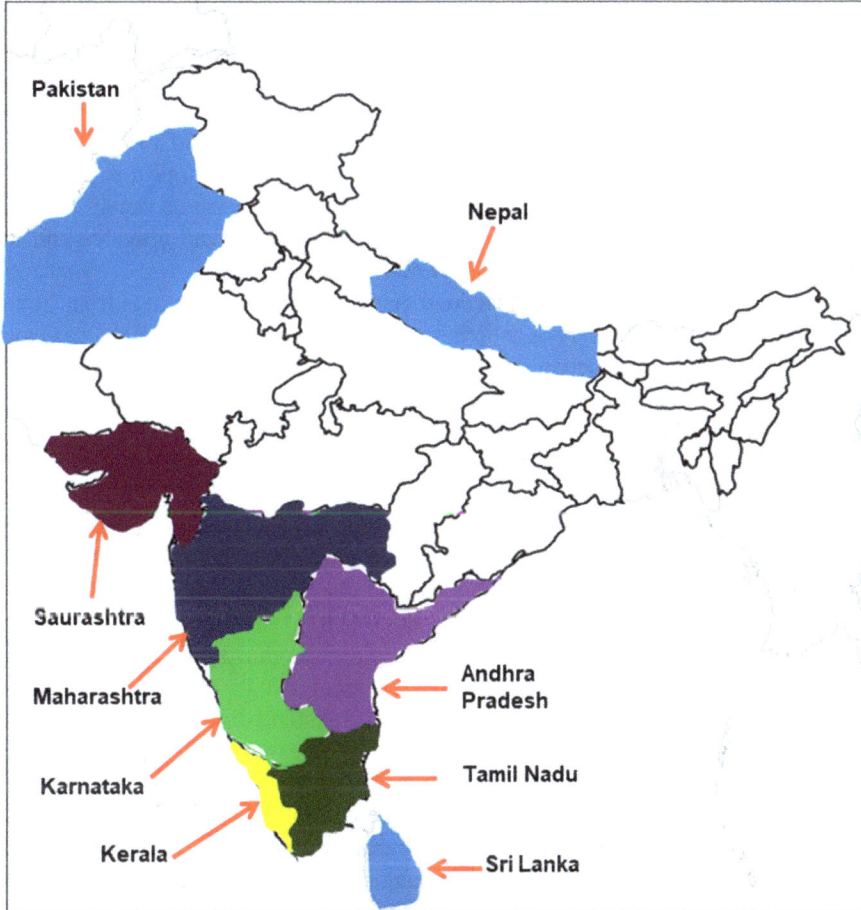

FIGURE 1.7 The Indian red scorpion's geographic range across the Indian subcontinent (blue fill: India's surrounding countries). (Image sketched by B. Das.)

2012; Pipelzadeh et al. 2007; Ghafourian et al. 2016; Dehghani and Fathi 2012; Olivier 1807; Yousef Mogaddam et al. 2017; Vazirianzadeh and Salahshoor 2015).

The nocturnal predators identified as Indian red scorpions are native to the Indian subcontinent (Figure 1.7). They are infrequently seen outside of Sri Lanka (Kularatne et al. 2015), Eastern Nepal (Bhadani et al. 2006), or Eastern Pakistan (Kovařík 2007). In India, the most common regions for reported rates of morbidity and death from scorpion stings include western Maharashtra, Saurashtra, Kerala, Andhra Pradesh, Tamil Nadu, and Karnataka. In India, regional heterogeneity in the intensity of scorpion stings has been documented (Reddy 2013; Suranse et al. 2019). This variance is most likely the outcome of genetic heterogeneity within the population, which causes phenotypic variations in venom composition (Newton et al. 2007).

Moderate genetic variation was found in several populations of Indian red scorpions that were collected from eight localities in western Maharashtra, India: Alandi, the Bhate plateau, Jalna, Jejuri, Kalyan, Sangameshwar, Pashan, and Shindavane. According to the regression study, the genetic distance between the subspecies rises with each kilometer of geographic separation by 0.006% (95% CI: 0.003–0.010%) (Suranse et al. 2017). Furthermore, it has been suggested that variances in rainfall—precisely, high, moderate, and low rainfall regions—correspond to variations in genetic makeup linked to changes in venom characteristics (Suranse et al. 2017). For instance, there was a noticeable difference in the production of venom peptides between Indian red scorpions taken from Maharashtra's Konkan province and the semi-arid Deccan plateau in West India (Newton et al. 2007).

Additionally, reports based on anecdotal evidence suggest that stings from Indian red scorpions in the Konkan area, which is on the western side of the Western Ghats, hurt more than stings from populations on the eastern side of the same region. These differences are probably caused by the different venom peptide compositions of the two populations (Newton et al. 2007). However, additional elements might also have a role in the pathophysiology of the sting (Newton et al. 2007; Kankonkar, Kulkurni, and Hulikavi 1998; Bawaskar and Bawaskar 1992).

Indian red scorpions from Southern India (Chennai) and Western India (Chiplun, Ratnagiri, and Ahmednagar) have also been shown to differ intra-specifically in their venom by using sodium dodecyl sulfate–polyacrylamide gel electrophoresis (SDS–PAGE) to analyze venom samples (Badhe et al. 2007). The blood sodium levels of mice given equivalent doses of Indian red scorpion venom from the aforementioned geographical regions varied significantly (Badhe et al. 2007). The plethora of toxins that target the central nervous and cardiovascular systems through the Na^+ and K^+ channels is responsible for the severe toxicity of the venom of Indian red scorpions (Das, Patra, and Mukherjee 2020; Das, Saviola, and Mukherjee 2021). The Iranian Scorpion (*H. lepturus*), a member of the Hemiscorpiidae family (Prendini 2000), and the endemic West African *Pandinus imperator*, a member of the Scorpionidae family, share the same venom profile. Although detailed examinations of the venoms of Indian red scorpions from various regions of the Indian subcontinent are presently absent, this information might assist in determining the influence of geography on the composition of venom. In comparison, it has been demonstrated that venom variations cause the severity of stings and symptoms in scorpions from different world locations (Abroug et al. 2020).

There are 2,584 species of scorpions today, divided into 23 families (Reckziegel and Pinto 2014; Laustsen et al. 2016; Lourenço 2018). Scorpion venom alone may not pose a medicinal threat to human health. Nevertheless, they may demonstrate a potential origin of bioenergetic particles that may aid in a drive to develop novel treatments against existing and emerging diseases.

Table 1.1 shows the distribution of medically significant scorpions around the world. The Buthidae family includes the most hazardous scorpion species that are lethal to humans (Figure 1.2) (Laustsen et al. 2016). Nevertheless, various types in the Scorpionidae and Hemiscorpidae families are also classified as mischievous (Lourenço 2018; Hauke and Herzig 2017).

The geographical distribution of prevalence of those medicinally important species is linked to the domestic spread of scorpions. Except for Antarctica, all continents are home to scorpions; however, the Amazon Basin region, the Middle East, southern

TABLE 1.1
Geographical distribution of scorpions around the world

S. No.	Scorpion Scientific Name	The Common Name of Scorpion	Family Name	Physiognomy/Characteristics	Geographical distribution		
					America	Asia	Africa
1	*Tityus serrulatus*	The yellow scorpion	Buthidae	They have a 5–7 cm length, serrated metasomal third and fourth segments, and parthenogenetic reproduction.	Brazil	-	-
2	*Tityus bahiensis*	The brown scorpion	Buthidae	They have a reddish-brown tail, dark trunk, legs, and palps with dark spots. The adult is around 7 cm long and exhibits sexual reproduction.	Brazil	-	-
3	*Tityus obscurus*	Amazonian black scorpion	Buthidae	Adults are black and grow to a maximum length of 9 cm. Young animals, however, are brown. They undergo sexual reproduction.	Brazil	-	-
4	*Tityus stigmurus*	The yellow scorpion	Buthidae	Regarding habits and coloration, they are similar to *T. serrulatus* but have a black longitudinal stripe in the dorsal region. Black adult specimens can grow to a height of 9 cm. They exhibit sexual activity for reproduction.	Brazil	-	
5	*Tityus pachyurus*		Buthidae	They range in length from 6 to 9 cm, have reddish pedipalps, and the fourth and fifth segments of the post-abdomen are thicker and darker.	Colombia	-	-

(continued)

TABLE 1.1 (Continued)
Geographical distribution of scorpions around the world

S. No.	Scorpion Scientific Name	The Common Name of Scorpion	Family Name	Physiognomy/Characteristics	Geographical distribution		
					America	Asia	Africa
6	*Tityus trivittatus*	Brazilian red house scorpion	Buthidae	They range in size from 50 to 70 mm, with a body color of yellow or reddish-yellow on the back, pale, yellow, or yellow-brown legs, pedipalps, and a tail that may also have dark segmented marks. The body color of the immature is reddish, with dark spots on the legs.	Argentina	-	
7	*Androctonus crassicauda*	The Arabian fat-tailed scorpion	Buthidae	Their shades range from light brown to reddish to black to blackish-brown. They can expand to more than 10 centimeters (3.9 in).	-	Iran, Turkey	North Africa
8	*Hemiscorpius lepturus*	—	Hemiscorpiidae	These scorpions have large bodies, long, thin tails, and a height of 8 cm for males and 5.5 cm for females.	-	Iran, Oman, Iraq, Saudi Arabia, Yemen, Pakistan	
9	*Scorpio maurus townsendi*	Large-clawed scorpion or Israeli gold scorpion	Scorpionidae	This Scorpionidae family member is a small to medium-sized scorpion that measures 3 inches (76 mm). Golden claws and a brown back are on it.	-	Iran	-
10	*Leiurus quinquestriatus*	Deathstalker	Buthidae	It measures 30–77 mm (1.2–3.0 in) in length, with an average of 58 mm, and is yellow.	-	-	Middle east and north Africa

No.	Scientific name	Common name	Family	Description		
11	*Mesobuthus tamulus*	Indian red scorpion	Buthidae	It is 50–90 mm in length. Telson granulated, and the tip of the pedipalps pincers are bright orange-yellow to light reddish-brown.	-	India, eastern Pakistan, eastern Nepal, Sri Lanka
12 njim	*Heterometrus bangalensis*	Indian black scorpion	Scorpionidae	It measures 95–115 mm in length. The body is dark reddish-brown to light brown.		West Bengal and Odisha
11	*Parabuthus granulatus*	Rough, thick tail scorpion	Buthidae	Its size is 11.5 cm, and its color ranges from dark yellow to brown. It has a reasonably tiny vesicle.	-	Sothern Africa
12	*Parabuthus transvaalicus*	Fattail scorpion	Buthidae	It is also known as the black thick-tailed scorpion because its tail is thick and dark brown or black in color and measures 90–110 mm (3.5–4.3 in) in length. Its tail is enlarged, with the sting segment as broad as the rest of the tail, but its pincers are thin.	-	Southern Africa
13	*Centruroides noxius*	—	Buthidae	The body of this species, which can reach lengths of 3.5–5 cm, is often black or brown, while its legs and pedipalps are typically white.	Western Mexico	-
14	*Centruroides sculpturatus*	Arizona Bark scorpion	Buthidae	An adult male scorpion can grow to a maximum length of 8 cm (3.14 in), whereas an adult female scorpion can grow to a full size of 7 cm (2.75 in).	Mexico	-

FIGURE 1.8 Global distribution of medically important scorpions (genus). (The regions are highlighted with a red circle.) (Photo Source: www.google.com/maps/@18.5882342,65.2452577,3z?entry=ttu; reprinted from the Google map.)

India, Brazil, the African Sahel, and South Africa have higher rates and severity of stings (Figure 1.8) (Santos et al. 2016; Ward, Ellsworth, and Nystrom 2018; Pezzi et al. 2019; Chippaux 2012). Nevertheless, few reports of symptoms brought on by the venom of most of these species exist (Ward, Ellsworth, and Nystrom 2018).

1.5   CONCLUSION

Arachnids, one of the most successful animal families on land, are still largely unknown about their evolutionary history beyond the sea. Therefore, constraining the phylogeny of the entire group of scorpions using fossil records and advances in phylogenetic divergence estimation is essential to address the evolutionary history of arachnids. For many decades, scorpions have confounded our knowledge of animal terrestrialization. Scorpions' diversity and distribution patterns are much more complex than they first appear. The fact that most evolutionary patterns are still unknown and the group has a relatively long evolutionary history makes this especially true. Although they are found primarily in tropical and subtropical parts, concerns about significant medical incidents involving scorpions are relevant worldwide.

REFERENCES

Abroug F, Ouanes-Besbes L, Tilouche N, and Elatrous S. 2020. Scorpion envenomation: state of the art. *Intensive Care Medicine* 46 (3):401–410.

Anderson EP, Schiffbauer JD, Jacquet SM, Lamsdell JC, Kluessendorf J, and Mikulic DG. 2021. Stranger than a scorpion: a reassessment of *Parioscorpio venator*, a problematic arthropod from the Llandoverian Waukesha Lagerstätte. *Palaeontology* 64 (3):429–474.

Badhe RV, Thomas AB, Deshpande AD, Salvi N, and Waghmare A. 2007. The action of red scorpion (*Mesobuthus tamulus* coconsis, Pocock) venom and its isolated protein fractions on blood sodium levels. *Journal of Venomous Animals Toxins including Tropical Diseases* 13:82–93.

Bawaskar HS, and Bawaskar PH. 1992. Management of the cardiovascular manifestations of poisoning by the Indian red scorpion (*Mesobuthus tamulus*). *Heart* 68 (11):478–480.

Bawaskar HS, and Bawaskar PH. 1998. Indian red scorpion envenoming. *The Indian Journal of Pediatrics* 65:383–391.

Bawaskar HS, and Bawaskar PH. 2012. Scorpion sting: update. *Journal of Association of Physicians of India* 60:46–55.

Bhadani U, Tripathi M, Sharma S, and Pandey R. 2006. Scorpion sting envenomation presenting with pulmonary edema in adults: a report of seven cases from Nepal. *Indian Journal of Medical Sciences* 60 (1):19–23.

Botton ML, Shuster Jr CN, Sekiguchi K, and Sugita H. 1996. Amplexus and mating behavior in the Japanese horseshoe crab, *Tachypleus tridentatus*. *Zoological Science* 13 (1):151–159.

Braddy SJ, and Dunlop JA. 2021. A sting in the tale of *Parioscorpio venator* from the Silurian of Wisconsin: is it a Cheloniellid arthropod? *Lethaia* 54 (5):603–609.

Bucaretchi F, Baracat EC, Nogueira RJ, Chaves A, Zambrone FA, Fonseca MR, and Tourinho FS. 1995. A comparative study of severe scorpion envenomation in children caused by *Tityus bahiensis* and *Tityus serrulatus*. *Revista do Instituto de Medicina Tropical de São Paulo* 37:331–336.

Bull EE, and Loydell DK. 1995. Uppermost Telychian graptolites from the North Esk Inlier, Pentland Hills, near Edinburgh. *Scottish Journal of Geology* 31 (2):163–170.

Casewell NR, Wüster W, Vonk FJ, Harrison RA, and Fry BG. 2013. Complex cocktails: the evolutionary novelty of venoms. *Trends in Ecology Evolution* 28 (4):219–229.

Chippaux JP. 2012. Emerging options for the management of scorpion stings. *Drug Design, Development*:165–173.

Coddington JA, Giribet G, Harvey MS, Prendini L, and Walter DE. 2004. Arachnida. In *Assembling the Tree of Life*, edited by J Cracraft, and MJ Donoghue. New York, NY: Oxford University Press:296–318.

Das B, Patra A, and Mukherjee AK. 2020. Correlation of venom toxinome composition of Indian red scorpion (*Mesobuthus tamulus*) with clinical manifestations of scorpion stings: failure of commercial antivenom to immune-recognize the abundance of low molecular mass toxins of this venom. *Journal of Proteome Research* 19 (4):1847–1856.

Das B, Saviola AJ, and Mukherjee AK. 2021. Biochemical and proteomic characterization, and pharmacological insights of Indian red scorpion venom toxins. *Frontiers in Pharmacology* 12:710680.

Dehghani R, and Fathi B. 2012. Scorpion sting in Iran: a review. *Toxicon* 60 (5):919–933.

Dehghani R, and Valayi N. 2005. Review on scorpions' taxonomy and Iranian scorpions' key identification. *Feiz* 32:73–92.

Dunlop JA. 2010. Geological history and phylogeny of Chelicerata. *Arthropod Structure Development* 39 (2–3):124–142.

Dunlop JA, Erik Tetlie O, and Prendini L. 2008. Reinterpretation of the Silurian scorpion *Proscorpius osborni* (Whitfield): integrating data from Palaeozoic and recent scorpions. *Palaeontology* 51 (2):303–320.

Dunlop JA, Scholtz G, and Selden PA. 2013. Water-to-land transitions. In *Arthropod Biology Evolution: Molecules, Development, Morphology*:417–439. Berlin, Heidelberg: Springer Berlin Heidelberg.

Dunlop JA, and Webster M. 1999. Fossil evidence, terrestrialization and arachnid phylogeny. *Journal of Arachnology* 2(1):86–93.

Eickstedt VRD, Ribeiro LA, Candido DM, Albuquerque MJ, and Jorge MT. 1996. Evolution of scorpionism by *Tityus bahiensis* (PERTY) and *Tityus serrulatus* Lutz and Mello and geographical distribution of the two species in the state of São Paulo–Brazil. *Journal of Venomous Animals and Toxins Including Tropical Diseases* 2 (2):92–105.

Farzanpay R. 1990. A catalogue of the scorpions occurring in Iran, up to January 1986. *Revue Arachnologique* 8 (2):1–12.

Fet V. 2000. *Suborder Branchioscorpionina Kjellesvig-Waering, 1986. Catalog of the Scorpions of the World*. New York: The New York Entomological Society:554–595.

Fet V, and Soleglad ME. 2005. Contributions to scorpion systematics. I. On recent changes in high-level taxonomy. *Euscorpius* 2005 (31):1–13.

Gess RW. 2013. The earliest record of terrestrial animals in Gondwana: a scorpion from the Famennian (Late Devonian) Witpoort Formation of South Africa: Arachnida. *African Invertebrates* 54 (2):373–379.

Ghafourian M, Ganjalikhanhakemi N, Hemmati AA, Dehghani R, and Kooti W. 2016. The effect of *Hemiscorpius lepturus* (Scorpionida: Hemiscorpiidae) venom on leukocytes and the leukocyte subgroups in peripheral blood of rat. *Journal of Arthropod-Borne Diseases* 10 (2):159.

Goyffon M, and Kovoor J. 1978. Chactoid venoms. Arthropod Venoms:395–418.

Hauke TJ, and Herzig V. 2017. Dangerous arachnids—fake news or reality? *Toxicon* 138:173–183.

Hjelle JT. 1990. Anatomy and morphology. In *The Biology of Scorpions*, edited by GA Polis. CABI:9–63.

Howard RJ, Edgecombe GD, Legg DA, Pisani D, and Lozano-Fernandez J. 2019. Exploring the evolution and terrestrialization of scorpions (Arachnida: Scorpiones) with rocks and clocks. *Organisms Diversity Evolution* 19:71–86.

Jeram AJ. 1997. Phylogeny, classification and evolution of Silurian and Devonian scorpions. Paper read at *Proceedings of the 17th European Colloquium of Arachnology, Edinburgh.*

Jones WT, Feldmann RM, and Schweitzer CE. 2015. Ceratiocaris from the Silurian Waukesha Biota, Wisconsin. *Journal of Paleontology* 89 (6):1007–1021.

Kankonkar RC, Kulkurni DG, and Hulikavi CB. 1998. Preparation of a potent anti-scorpion-venom-serum against the venom of red scorpion (*Buthus tamalus*). *Journal of Postgraduate Medicine* 44 (4):85.

Kleffner MA, Norby RD, Kluessendorf J, and Mikulic DG. 2018. Revised conodont biostratigraphy of Lower Silurian strata of southeastern Wisconsin. Paper read at *GSA North-Central 2018 Annual Meeting.*

Kovařík F. 2007. A revision of the genus *Hottentotta Birula*, 1908, with descriptions of four new species (Scorpiones, Buthidae). *Euscorpius* 2007 (58):1–107.

Kühl G, Bergmann A, Dunlop J, Garwood RJ, and Rust JE. 2012. Redescription and palaeobiology of *Palaeoscorpius devonicus* Lehmann, 1944 from the Lower Devonian Hunsrück slate of Germany. *Palaeontology* 55 (4):775–787

Kularatne SA, Dinamithra NP, Sivansuthan S, Weerakoon KG, Thillaimpalam B, Kalyanasundram V, and Ranawana, KB. 2015. Clinico-epidemiology of stings and envenoming of *Hottentotta tamulus* (Scorpiones: Buthidae), the Indian red scorpion from Jaffna Peninsula in northern Sri Lanka. *Toxicon* 93:85–89.

Lamoral BH. 1980. A reappraisal of the suprageneric classification of recent scorpions and of their zoogeography. Wien: Proceedings of the 8th International Archaeological Congress:439.

Laurie, M. 1900. XIX.—On a Silurian Scorpion and some additional Eurypterid Remains from the Pentland Hills. *Earth Environmental Science Transactions of The Royal Society of Edinburgh* 39 (3):575–590.

Laustsen AH, Solà M, Jappe EC, Oscoz S, Lauridsen LP, and Engmark M. 2016. Biotechnological trends in spider and scorpion antivenom development. *Toxins* 8 (8):226.

Legg DA, Garwood RJ, Dunlop JA, and Sutton M. 2012. A taxonomic revision of orthosternous scorpions from the English Coal-Measures aided by X-ray micro-tomography (XMT). *Palaeontologia Electronica* 15 (2):1–16.

Lira-da-Silva RM, Amorim AMD, and Brazil TK. 2000. Envenenamento por *Tityus stigmurus* (scorpiones; Buthidae) no estado da Bahia, Brasil. *Revista da Sociedade Brasileira de Medicina Tropical* 33:239–245.

Loret E, Hammock BD, Brownell P, and Polis GA. 2001. *Scorpion Biology and Research*, edited by Brownell P, and Polis GA. New York: Oxford University Press, Inc.

Lourenço WR. 1985. *Essai d'interprétation de la distribution du genre Opisthacanthus (Arachnida, Scorpiones, Ischnuridae) dans les régions néotropicale et afrotropicale.* Museum National d'Histoire Naturelle, Labo. de Zool. (Arthropodes).

Lourenço WR. 2016. Scorpion incidents, misidentification cases and possible implications for the final interpretation of results. *Journal of Venomous Animals and Toxins including Tropical Diseases* 22:21.

Lourenço WR. 2018. The evolution and distribution of noxious species of scorpions (Arachnida: Scorpiones). *Journal of Venomous Animals Toxins Including Tropical Diseases* 24.

Lourenço WR, and Gall J. 2004. Fossil scorpions from the Buntsandstein (early Triassic) of France. *Comptes Rendus Palevol* 3 (5):369–378.

Loydell DK. 2005. Graptolites from the Deerhope Formation, North Esk Inlier. *Scottish Journal of Geology* 41 (2):189–190.

Magnani F, Stockar R, and Lourenço WR. 2022. A new family, genus and species of fossil scorpion from the Meride Limestone (Middle Triassic) of Monte San Giorgio (Switzerland). *Faunitaxys* 10 (24):1–7.

McCoy VE, and Brandt DS. 2009. Scorpion taphonomy: criteria for distinguishing fossil scorpion molts and carcasses. *The Journal of Arachnology* 37 (3):312–320.

Mikulic DG, Briggs DE, and Kluessendorf J. 1985. A Silurian soft-bodied biota. *Science* 228 (4700):715–717.

Mullen GR, and Sissom WD. 2019. Scorpions (scorpiones). In *Medical and Veterinary Entomology*, edited by Gary R. Mulen and Lance A. Durden: Academic Press.

Newton KA, Clench MR, Deshmukh R, Jeyaseelan K, and Strong PN. 2007. Mass fingerprinting of toxic fractions from the venom of the Indian red scorpion, *Mesobuthus tamulus*: biotope-specific variation in the expression of venom peptides. *Rapid Communications in Mass Spectrometry* 21 (21):3467–3476.

Nishikawa AK, Caricati CP, Lima ML, Dos Santos MC, Kipnis TL, Eickstedt VR, Knysak I, Da Silva MH, Higashi HG, and Da Silva WD. 1994. Antigenic cross-reactivity among the venoms from several species of Brazilian scorpions. *Toxicon* 32 (8):989–998.

Olivier GA. 1807. *Voyage dans l'Empire Othoman, l'Egypte et la Perse: fait par ordre du gouvernement, pendant les six premières années de la République: avec atlas.* Vol. 3. Agasse.

Petrunkevitch A. 1953. *Paleozoic and Mesozoic Arachnida of Europe.* Vol. 53. Geological Society of America.

Pezzi M, Bonacci T, Leis M, Mamolini E, Marchetti MG, Krčmar S, Chicca M, Del Zingaro CN, Faucheux MJ, and Scapoli C. 2019. Myiasis in domestic cats: a global review. *Parasites Vectors* 12 (1):1–14.

Pipelzadeh MH, Jalali A, Taraz M, Pourabbas R, and Zaremirakabadi A. 2007. An epidemiological and a clinical study on scorpionism by the Iranian scorpion *Hemiscorpius lepturus*. *Toxicon* 50 (7):984–992.

Porto TJ, Brazil TK, and Lira-da-Silva RM. 2010. Scorpions, state of Bahia, northeastern Brazil. *Check List* 6 (2):292–297.

Poschmann M, Dunlop JA, Kamenz C, and Scholtz G. 2008. The Lower Devonian scorpion Waeringoscorpio and the respiratory nature of its filamentous structures, with the description of a new species from the Westerwald area, Germany. *Paläontologische Zeitschrift* 82:418–436.

Prendini L. 2000. Phylogeny and classification of the superfamily Scorpionoidea Latreille 1802 (Chelicerata, Scorpiones): an exemplar approach. *Cladistics* 16 (1):1–78.

Prendini L, and Linder HP. 1998. Phylogeny of the South African species of restioid leafhoppers, tribe Cephalelini (Homoptera: Cicadellidae, Ulopinae). *Insect Systematics Evolution* 29 (1):11–18.

Prendini L, and Wheeler WC. 2005. Scorpion higher phylogeny and classification, taxonomic anarchy, and standards for peer review in online publishing. *Cladistics* 21 (5):446–494.

Reckziegel GC, and Pinto VL. 2014. Scorpionism in Brazil in the years 2000 to 2012. *Journal of Venomous Animals Toxins Including Tropical Diseases* 20:2–8.

Reddy BRC. 2013. Scorpion Envenomation: What Is New? Medicine Update. *Nova Delhi, India: Jaypee Brothers Medical Publishers*: 421–423.

Rein JO. *The Scorpion Files.* Norwegian University of Science and Technology, Trondheim. 2020.

Rubio M. 2000. Commonly available scorpions. In *Scorpions: Everything about Purchase, Care, Feeding, and Housing.* Hauppauge, NY: Barron's.

Santos MS, Silva CG, Neto BS, Grangeiro Júnior CR, Lopes VH, Teixeira Júnior AG, Bezerra DA, Luna JV, Cordeiro JB, Júnior JG, and Lima MA. 2016. Clinical and epidemiological aspects of scorpionism in the world: a systematic review. *Wilderness Environmental Medicine* 27 (4):504–518.

Saúde M. 2001. Manual de diagnóstico e tratamento de acidentes por animais peçonhentos. In *Fundação Nacional de Saúde/Coordenação de Controle de Zoonoses e Animais Peçonhentos*. CENEPI: Brasília.

Schoenemann B, Poschmann M, and Clarkson EN. 2019. Insights into the 400 million-year-old eyes of giant sea scorpions (Eurypterida) suggest the structure of Palaeozoic compound eyes. *Scientific Reports* 9 (1):17797.

Scholtz G, and Kamenz C. 2006. The book lungs of Scorpiones and Tetrapulmonata (Chelicerata, Arachnida): evidence for homology and a single terrestrialisation event of a common arachnid ancestor. *Zoology* 109 (1):2–13.

Sedaghat MM, and Moghadam ARS. 2012. Mapping the distribution of some important scorpions collected in the past five decades in Iran. *Annals of Military and Health Sciences Research* 9 (4):285–296.

Selden PA, Dunlop JA, and Edgecombe GD. 1998. Fossil taxa and relationships of chelicerates. *Arthropod Fossils Phylogeny* 303–331.

Selden PA, and Jeram AJ. 1989. Palaeophysiology of terrestrialisation in the Chelicerata. *Earth Environmental Science Transactions of The Royal Society of Edinburgh* 80 (3–4):303–310.

Sharma PP, Fernández R, Esposito LA, González-Santillán E, and Monod L. 2015. Phylogenomic resolution of scorpions reveals multilevel discordance with morphological phylogenetic signal. *Proceedings of the Royal Society B: Biological Sciences* 282 (1804):20142953.

Shorthouse DJ, and Marples TG. 1982. The life stages and population dynamics of an arid zone scorpion *Urodacus yaschenkoi* (Birula 1903). *Australian Journal of Ecology* 7 (2):109–118.

Sissom WD. 1990. Systematics, biogeography and paleontology. In The Biology of Scorpions, edited by GA Polis. Vol. 65. Stanford: Stanford University Press:64–160.

Soleglad ME, and Fet V. 2003. High-level systematics and phylogeny of the extant scorpions (Scorpiones: Orthosterni). *Euscorpius* 2003 (11):1–56.

Sridhara S, Chakravarthy AK, Kalarani V, and Reddy DC. 2016. Diversity and ecology of scorpions: evolutionary success through venom. In *Arthropod Diversity Conservation in the Tropics Sub-Tropics*. Springer:57–80.

Stockwell SA. 1989. *Revision of the Phylogeny and Higher Classification of Scorpions (Chelicerata)*. Berkeley: University of California, ProQuest Dissertations Publishing.

Suranse V, Sawant NS, Bastawade DB, Dahanukar N. 2019. Haplotype diversity in medically important red scorpion (Scorpiones: Buthidae: *Hottentotta tamulus*) from India. *Journal of Genetics* 98 (1):17.

Suranse V, Sawant NS, Paripatyadar SV, Krutha K, Paingankar MS, Padhye AD, Bastawade DB, and Dahanukar N. 2017. First molecular phylogeny of scorpions of the family Buthidae from India. *Mitochondrial DNA Part A* 28 (4):606–611.

Vazirianzadeh B, and Salahshoor A. 2015. Scorpion sting in Izeh, Iran: an epidemiological study during 2009–2011. *Journal of Basic Applied Sciences* 11:403.

Waddington J, Rudkin DM, and Dunlop JA. 2015. A new mid-Silurian aquatic scorpion—one step closer to land? *Biology Letters* 11 (1):20140815.

Ward MJ, Ellsworth SA, and Nystrom GS. 2018. A global accounting of medically significant scorpions: epidemiology, major toxins, and comparative resources in harmless counterparts. *Toxicon* 151:137–155.

Wendruff AJ, Babcock LE, Kluessendorf J, and Mikulic DG. 2016. The Waukesha biota: an unusual glimpse of life on a Silurian carbonate platform. *Geological Society of America Abstracts with Programs* 48 (5). doi: 10.1130/abs/2016NC-275201

Wendruff AJ, Babcock LE, Wirkner CS, Kluessendorf J, and Mikulic DG. 2020. A Silurian ancestral scorpion with fossilised internal anatomy illustrating a pathway to arachnid terrestrialisation. *Scientific Reports* 10 (1):14.

Yang S, Yang F, Zhang B, Lee BH, Li B, Luo L, Zheng J, and Lai R. 2017. A bimodal activation mechanism underlies scorpion toxin–induced pain. *Science advances* 3 (8):e1700810.

Yousef Mogaddam M, Dehghani R, Enayati AA, Fazeli-Dinan M, Vazirianzadeh B, Yazdani-Cherati J, and Motevalli Haghi F. 2017. Scorpion fauna (Arachnida: Scorpiones) in Darmian County, Iran (2015–2016). *Journal of Mazandaran University of Medical Sciences* 26 (144):108–118.

2 Scorpions
Taxonomic Classification, Anatomy, Morphology, Habitat, and Reproduction

2.1 INTRODUCTION

The term "scorpion" refers to the predatory arachnids of the order Scorpiones. They have eight legs, and their slender, segmented tail and pair of gripping pincers help to identify them. The tail frequently curves forward over the back and always has a stinger at the end (Bücherl 1971). Scorpions have undergone evolutionary change for 435 million years. They are present on all continents except Antarctica but live primarily in deserts and have evolved in various environmental conditions (Lourenço 2018). Of the more than 2,500 identified species, 22 recognized extant (living) families exist today.

Although some species of scorpions pursue vertebrates, most feed mainly on insects and other invertebrates (Howard et al. 2019). They use their pincers to catch and kill prey or to prevent themselves from a predator. The deadly sting is employed both offensively and defensively. Most species of scorpions do not pose any particular threat to people; healthy individuals usually do not require medical care after being stung. Only 25 species (less than 1% of total scorpion species) have venom that can kill a human, and this happens regularly in the areas of the world where access to medical care is meager (Lourenço et al. 1996).

Scorpions have been depicted in mythology, folklore, commercial branding, and the visual arts. To avoid being stung, Kilim carpets have scorpion themes. The constellation Scorpius is named after the astrological sign Scorpio. A traditional Scorpius story describes how the enormous scorpion and its adversary, Orion, evolved into constellations on either side of the sky (Kerényi 1974).

2.2 TAXONOMIC CLASSIFICATION OF SCORPIONS

Between 1758 and 1767, Carl Linnaeus named six species of scorpions under the name Scorpio; three are today recognized as valid species under the name *Androctonus australis*, *Euscorpius carpathicus*, and *Scorpio maurus*, while the characteristics of the other three are dubious. The scorpions were added to this "Insecta aptera" (group of wingless insects), which also contained Crustacea, Arachnida, and Myriapoda (Fet, Braunwalder, and Cameron 2002).

DOI: 10.1201/9781003540816-2

Jean-Baptiste Lamarck divided the "Insecta Aptera" in 1801, giving rise to the taxon Arachnides. They included Thysanura (thrips), Myriapoda, and parasites like lice, in addition to spiders, scorpions, and acari (mites and ticks) (Burmeister 1836). The Scorpiones were established in 1837 by Carl Ludwig Koch, a German arachnologist. He divided them up into four families: "Scorpionides" (scorpions with 6 eyes), "Buthides" (scorpions with 8 eyes), "Centrurides" (scorpions with 10 eyes), and "Androctonides" (scorpions with 12 eyes) (Koch 1837). Since then, 22 families and some 2,500 species of scorpions have been discovered and described; in the 21st century, multiple taxonomic additions and reorganizations happened (Howard et al. 2019). More than 100 taxa of prehistoric scorpions are described in the literature (Dunlop and Penney, 2012). This classification is based on Soleglad and Fet's (2003) work, which replaced Stockwell's (1989) earlier unpublished work. The studies of Soleglad and colleagues contain more taxonomic modifications (Soleglad, Fet, and Kovařík 2005; Fet and Soleglad 2005).

The taxonomic classification of scorpion
Kingdom: Animalia
Phylum: Arthropoda
Subphylum: Chelicerata
Order: Scorpiones
Family: a. Buthidae,
b. Hemiscorpionidae,
c. Scorpionidae
Genus: a. *Tityus, Androctonus, Leiurus, Parabuthus, Centruroides*
b. *Hemiscorpius*
c. *Scorpio*

2.3   ANATOMY AND MORPHOLOGY

The *Typhlochactas mitchelli* of the Typhlochactidae is 8.5 mm (0.33 in) in size, while the *Heterometrus swammerdami* of the Scorpionidae is 23 cm (9.1 in) in length (Rubio 2000). The cephalothorax, also known as the prosoma, and the abdomen, often called the opisthosoma, are the two tagmata that form the body of a scorpion. The opisthosoma is divided into two parts: The sizeable anterior mesosoma, the preabdomen, and the small, tail-like posterior metasoma, often known as the postabdomen (Polis 1990). In most animals, there are no obvious external differences between the sexes. However, in rare cases, male metasomae are more extended than females (Stockmann, Ythier, and Fet 2010). The body of a scorpion is composed of three main sections: (i) cephalothorax, (ii) mesosoma, and (iii) metasoma (Klußmann-Fricke, Prendini, and Wirkner 2012; Lourenço 2020) (Figure 2.1).

2.3.1   Prosoma or Cephalothorax

The carapace, eyes, chelicerae (mouth parts), pedipalps (which feature chelae, also called claws or pincers), and four pairs of walking legs comprise the cephalothorax. Two pairs of eyes are located on top of the cephalothorax of scorpions, while two to five

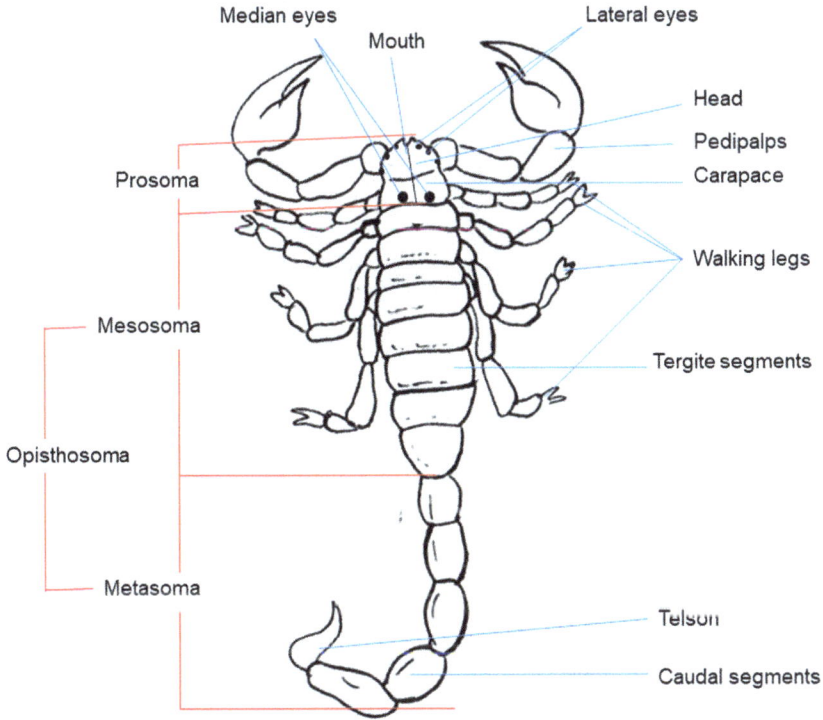

FIGURE 2.1 Anatomy of scorpion. (Image sketched by B. Das.)

pairs are typically found in the front corners of the cephalothorax. As its primary eyes are the most light-sensitive in the animal kingdom, especially in low light, nocturnal species may navigate at night using starlight even though their eyes cannot produce precise images. The front and underside of the carapace are home to the chelicerae. They resemble tongs and contain three segments and sharp "teeth" (Sridhara et al. 2016). The scorpion's brain is behind the cephalothorax, just above the esophagus (Figure 2.1).

In scorpions, the nervous system is concentrated primarily in the cephalothorax, as in other arachnids. However, it also has segmented ganglia and a long ventral nerve cord, which may be primitive features. The segmented, clawed pedipalp is an appendage used for protection, immobilizing prey, and sensory purposes. The coxa, trochanter, femur, patella, tibia (which includes the manus and fixed claw), and tarsus make up the pedipalp. These parts are listed from closest to the body outward (moveable claw). The pedipalp segments and other portions of the scorpion's body have darker or granular elevated linear ridges known as "keels" or "carinae," which are essential taxonomic characteristics (Coelho et al. 2017). Unlike other arachnids, their legs have not evolved for any other use; however, they may occasionally be used for digging, and females may use them to capture their emerging young. Proprioceptors, bristles, and sensory setae are present throughout the legs. Depending on the species of scorpions, spurs and spines may be present on their legs (Figure 2.2) (Coelho et al. 2017; Farley 2011).

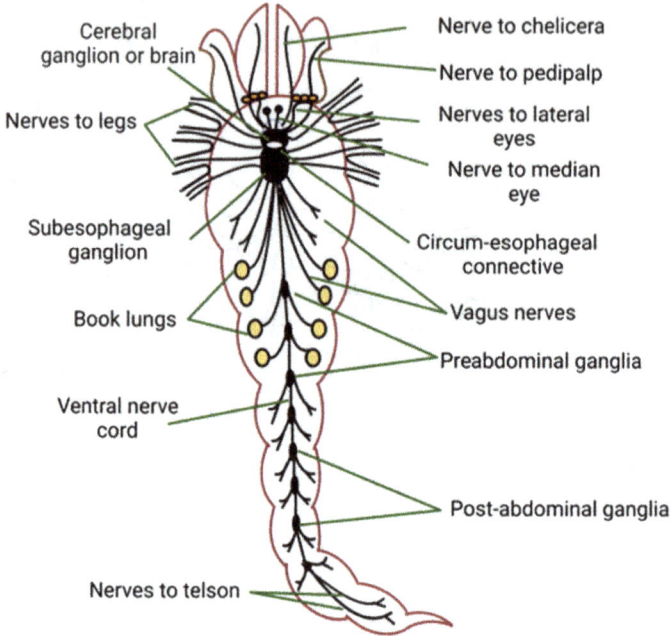

FIGURE 2.2 Nervous system of a scorpion. (Created with BioRender.com; agreement number: XG25HMISCM.)

2.3.2 PREABDOMEN OR MESOSOMA

The broad portion of the opisthosoma is the mesosoma or preabdomen. Eight segments comprise the mesosoma during the early stages of embryonic development. However, the first section disappears before birth in scorpions, leaving only segments 2–8 (Polis 1990). A sclerotized plate of scorpion, known as the tergite, covers each of the opisthosoma's anterior seven somites (segments) on the dorsal side. Somites 3–7 are defended ventrally by sternites, which are matched plates. Two genital opercula cover the gonopore on the ventral side of somite 1. The basal plate consists of sternite 2 and contains the pectines, which serve as sensory organs (Knowlton and Gaffin 2011). The following four somites, 3–6, have two spiracles (Figure 2.1). These act as book lungs—the respiratory organs in scorpions used for atmospheric gas exchange.

Depending on the species, the spiracle openings might be oval, circular, elliptical, or slit-shaped (Polis 1990). Hence, there are four sets of book lungs. Each group has 140–150 air-filled pulmonary chambers connected on the ventral side to atrial chambers that open into spiracle-like structures. The lamellae are kept separated by bristles. The dorsoventral muscles compress the pulmonary compartment during contraction, which forces air out. The muscles subsequently relax to allow the chamber to fill again. A muscle extends the atrial chamber and opens the spiracle (Polis 1990). The seventh and final somite lacks notable exterior features or appendages (Wanninger 2015).

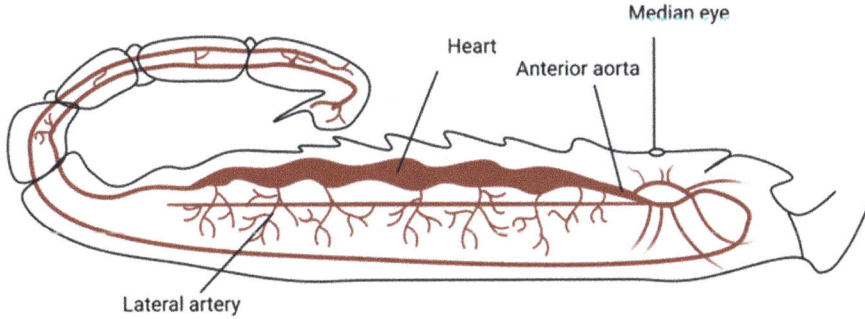

FIGURE 2.3   Circulatory system of a scorpion is mainly controlled by the heart. (Created with BioRender.com; agreement number: QJ25HMJ3NH.)

The scorpion's open circulatory system is controlled by the heart, also known as the "dorsal vessel," in the mesosoma. A deep artery network that runs throughout the body is continuous with the heart. Deoxygenated blood (hemolymph), which is returned to the heart by the sinuses, is reoxygenated by cardiac pores (Figure 2.3).

The reproductive system is also located within the mesosoma. The female gonads have two to four transverse anastomoses that connect three or four parallel tubes. Both egg production and embryo development take place in these tubes. They connect to two oviducts, which relate to one atrium and the genital pore, respectively (Stockmann 2015). Two gonads are cylindrical tubes arranged ladder-like and contain cysts that generate spermatozoa. One on either side of the mesosoma, the two tubes converge in spermiducts. They connect to the paraxial organs, symmetrical glandular structures at the genital opening. They form chitin structures that together form spermatophores (Stockmann 2015; Lautie et al. 2008).

2.3.3   TAIL OR METASOMA

Five segments and the telson compensate for the "tail" or metasoma, which is not a part. Body rings make up the five segments. They tend to be bigger distally and lack an apparent sterna or terga. These portions can be categorized based on taxonomy because of their bristles, setae, and keels. Four anal papillae and the anal arch surround the anus at the final segment's distal and ventral end (Polis 1990). Several scorpions have light receptors in their tails (Sridhara et al. 2016). The vesicle is a component of the telson and houses a symmetrical pair of venom glands. Its curved, sensory-haired hypodermic aculeus stinger is visible on the outside. Each venom gland has a separate duct that carries its secretion from the bulb of the gland, an area immediately adjacent to the tip, where each paired duct has a distinct venom pore (Yigit and Benli 2010).

Comparatively, a muscle system inside the glands pumps venom into the selected victim through the stinger, while a muscle system outside the glands in the tail propels. It penetrates the aculeus (Stockmann 2015). Zinc-rich metalloproteins in the stinger stiffen the tip (Schofield 2001). An angle of about 30 degrees relative to the tip is ideal for stinging (van der Meijden et al. 2013).

2.4   HABITAT OF SCORPIONS

The majority of scorpion species are crepuscular or nocturnal, spending their days hidden under rock crevices, tree bark, and burrows (Stockmann 2015). Many species make homes under stones a few millimeters long. Some may use the tunnels dug by small mammals, reptiles, spiders, and other creatures. Some animals make tunnels of varying complexity and depth. Burrows about 2 m (6 ft 7 in) deep are dug by *Hadrurus* species. The legs, claws, and mouth parts are used for digging. Individuals may congregate in the same shelter in several species, especially those belonging to the Buthidae family; bark scorpions can aggregate in groups of up to 30 individuals. Families of females and young may occasionally congregate together in species (Stockmann, Ythier, and Fet 2010). Scorpions can resist extreme heat: If adequately hydrated, *Leiurus quinquestriatus*, *S. maurus*, and *Hadrurus arizonensis* may survive at 45–50°C (113–122°F). Scorpions prefer environments with temperatures ranging from 11 to 40°C (52–104°F).

Desert species have to adapt to the drastic changes in temperature from day to night or from season to season. *Pectinibuthus birulai*, a psammophile scorpion, can tolerate temperatures between 30 and 50°C (22 and 122°F). Scorpions living outside the desert prefer cooler temperatures (Stockmann 2015).

Desert scorpions have many adaptations to conserve water. Without water, they eliminate insoluble substances from the body, such as xanthine, guanine, and uric acid. The primary component, guanine, enhances nitrogen excretion. Using lipids and waxes produced by epidermal glands, a scorpion's cuticle traps moisture and provides ultraviolet (UV) protection. A dehydrated scorpion can tolerate high blood osmotic pressure (Cowles 2018). While some desert scorpions can take in water from the moist soil, most obtain moisture from their food. Animals prefer cooler climates and dense vegetation and drink water from puddles and plants (Stockmann, Ythier, and Fet 2010). The scorpion's sting is used for both offense and defense. Others deliver slower, circular attacks to make it easier to reposition the stinger so it can sting again. Some animal species use their tails to strike hard and fast. During a defensive strike, *Leiurus quinquestriatus* can slap its tail at up to 128 cm/s (50 in/s) (Coelho et al. 2017).

Some other arthropods, such as centipedes, ants, spiders, and solifugids, may attack scorpions. The primary predators are birds, mammals, lizards, snakes, and frogs (Stockmann, Ythier, and Fet 2010). Meerkats are adept at catching and eating scorpions because they can bite off their stingers and are immune to venom. The desert long-eared bat and grasshopper mouse are other predators that have evolved to hunt scorpions and are resistant to their venom (Thompson 2018; Holderied, Korine, and Moritz 2011).

One investigation found scorpion particles in 70% of the latter's excretions (Holderied, Korine, and Moritz 2011). Scorpions host various parasites, including bacteria, nematodes, mites, and scuttle flies. Due to their immune systems, scorpions resist many bacterial infections (Stockmann 2015). When attacked by an object, the scorpion takes a defensive stance by raising its claws and tail. Certain animals stridulate, rubbing their stingers or claws together, to scare off predators (Stockmann, Ythier, and Fet 2010). Depending on the size of the appendages, certain species choose to defend themselves using claws or stingers (van der Meijden et al. 2013). Some scorpions, including *Parabuthus granulatus*, *Centruroides margaritatus*, and *H. arizonensis*, have been observed to spew venom in a narrow stream as far as 1 m (3.3 ft) to scare off potential predators and possibly cause eye damage (Stockmann, Ythier, and Fet 2010). Several *Ananteris* species can shed portions of their tails to avoid predators. They cannot sting or poop because the parts do not grow back, but they can continue to trap small prey and spawn for at least 8 months after that (Mattoni et al. 2015).

2.5 FOOD OF SCORPIONS

Scorpions feed on insects, especially wasps, grasshoppers, crickets, termites, and beetles. Other prey include spiders, woodlice, snakes, lizards, and even small vertebrates like mammals (Figure 2.4). Earthworms and mollusks may be prey for large-clawed species. *Isometroides vescus* is a species that only eats burrowing spiders; however, the rest are opportunistic and eat a variety of prey (Figure 2.4). The species' size influences the prey's size. Several species of scorpions are sit-and-wait predators that wait for food at or close to the entryway to their tunnel. Others deliberately seek them out. Scorpions use the mechanoreceptive and chemoreceptive hairs on their body to locate their prey and then use their claws to trap it. Particularly in the case of species with large claws, smaller animals are destroyed by the claws. Prey that is larger and more aggressive is stung (Stockmann 2015). Similar to other arachnids, scorpions digest their food externally.

A small amount of food is pulled off the prey item and placed in a preoral compartment beneath the chelicerae and carapace using extremely sharp chelicerae. Food is digested with the digestive juices from the gut, after which the liquid digestion of the food is sucked into the stomach. Setae catch and expel any solid, indigestible material (exoskeleton fragments) from the preoral cavity. The pharynx pushes the food sucked into the midgut, where it is further broken down. The anus and the hindgut are the

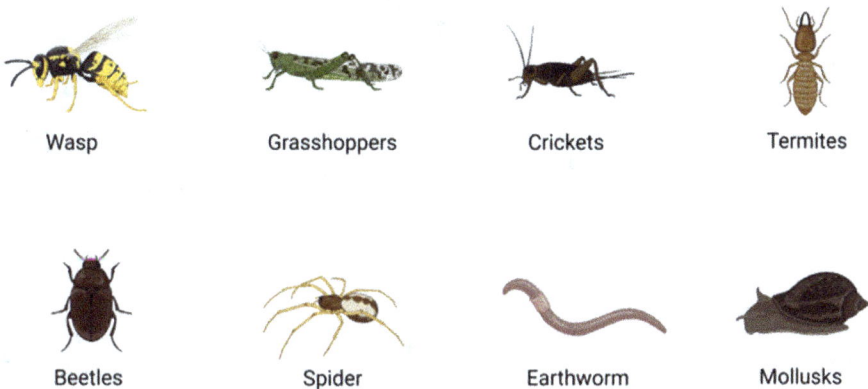

| Wasp | Grasshoppers | Crickets | Termites |

| Beetles | Spider | Earthworm | Mollusks |

FIGURE 2.4 Foods of scorpion. (Created with BioRender.com; agreement number: JF25HMJNPF.)

places where the waste passes out. Scorpions have enormous appetites and can eat a lot of food at once. They lead somewhat sedentary lives and have highly efficient organs for storing food (Polis 1990).

2.6 REPRODUCTION IN SCORPIONS

Most scorpions reproduce sexually. However, some genera, such as *Hottentotta* and *Tityus*, as well as the species *Ananteris coineaui, Centruroides gracilis*, and *Liocheles australasiae*, have recorded cases of parthenogenesis, the process by which unfertilized eggs grow into viable embryos (Lourenço 2008). Males on the move gather up pheromones from receptive females using their pectines to comb the substrate. Juddering is the back-and-forth body movement that males use when they begin courting. The female appears to detect ground vibrations due to this (Stockmann 2015). The two then do a dance known as the promenade à deux by making contact with their pedipalps (French for "a walk for two"). While the male looks for an excellent spot to deposit his spermatophore, the female and he dances back and forth, facing each other (Figure 2.5). The arbre droit (French for "upright tree") behavior, in which partners lift and touch their rears together, cheliceral kissing, in which the male and female hold each other's mouth parts, and sexual stinging, in which the male stings the female in the chelae or mesosoma to dominate her, are additional behaviors that may be a part of the courtship ritual. The dance could last a few minutes or several hours (Stockmann 2015; Stockmann, Ythier, and Fet 2010).

FIGURE 2.5 Scorpions in a courting dance. (Photo source: https://upload.wikimedia.org/wikipedia/commons/6/64/Dancing_scorpions-66970ep.jpg.) (Photo courtesy: prof.bizzarro; CC-BY 2.0.)

In some species of scorpions, the gestation period can last up to 1 year (Polis 1990). Both apoikogenic and katoikogenic embryonic development are possible. In the apoikogenic system, mainly in the Buthidae family, embryos develop in yolk-rich eggs inside follicles. In the katoikogenic system, which has been observed in Hemiscorpiidae, Scorpionidae, and Diplocentridae, the embryos grow in a diverticulum with a teat-like structure through which they can feed (Warburg 2010). Scorpions appear universally viviparous, giving birth to live young, unlike most arachnids, which are oviparous and hatch from eggs (Warburg 2012). The attention given by a mother to her young distinguishes them from other terrestrial arthropods (Polis 1990). Depending on the species, the size of a brood can range from 3 to 100 (Lourenço 2020). Neither the brood's size nor the life cycle's duration is linked to the scorpion's body size (Monge-Nájera 2019).

The female raises her front end before giving birth, placing her pedipalps and front legs underneath her to collect the young ("birth basket") (Figure 2.6). One by one, the young emerge from the genital opercula, eject any potential embryonic membrane, and settle upon their mother's back, where they stay until they reach juvenile status and have gone through minimum one molt. The pro-juvenile stage is before the first molt; during this stage, the young cannot feed or sting but have suckers on their tarsi to cling to their mother. This period can vary from 5 to 25 days, contingent upon the species. The child's first synchronized molt marks adolescence, which takes 6–8 hours (Lourenço 2020).

FIGURE 2.6 Female scorpion with young. (Photo source: https://upload.wikimedia.org/wikipedia/commons/0/00/Arizona_bark_scorpion_mom_and_babies.JPG.) (Photo courtesy: David S. Flores; CC BY-SA 3.0.)

The instars or juvenile stages resemble their smaller counterparts as adults and are fully equipped with pincers, hairs, and stings. They continue to ride on their mother's back for safety because they are still delicate and lack color (Figure 2.6). Over the next few days, they will become more intricate and colorful. If they perceive any danger, they may separate from their mother for some time before returning. Upon adulthood and the exoskeleton hardening, the young may hunt prey independently and eventually leave the mother (Stockmann, 2015). Before maturing, a scorpion may molt an average of six times. This process could take 6–83 months. Many creatures have a lifespan of 25 years (Polis 1990).

REFERENCES

Bücherl W. 1971. Classification, biology and venom extraction of scorpions. *Venomous Animals and their Venoms. Venomous Invertebrates* 3:317–347.

Bull EE, and Loydell DK. 1995. Uppermost Telychian graptolites from the North Esk Inlier, Pentland Hills, near Edinburgh. *Scottish Journal of Geology* 31 (2):163–170.

Bullock S, and Manias E. 2013. *Fundamentals of Pharmacology*. Pearson Higher Education Australia.

Burmeister H. 1836. *A Manual of Entomology*, translated by WE Shuckard. London: Edward Churton.

Cowles J. 2018. *Amazing Arachnids*. United States: Princeton University Press: ISBN 9780691176581.

Dunlop JA, and Penney D. 2012. *Fossil Arachnids*. United Kingdom: Siri Scientific Press. ISBN 0956779549.

Farley RD. 2011. The ultrastructure of book lung development in the bark scorpion *Centruroides gracilis* (Scorpiones: Buthidae). *Frontiers in Zoology* 8 (1):1–24.

Fet V, Braunwalder ME, and Cameron HD. 2002. Scorpions (Arachnida, Scorpiones) described by Linnaeus. *Bulletin-British Arachnological Society* 12 (4):176–182.

Fet V, and Soleglad ME. 2005. Contributions to scorpion systematics. I. On recent changes in high-level taxonomy. *Euscorpius* 2005 (31):1–13.

Holderied M, Korine C, Moritz T. 2011. Hemprich's long-eared bat (*Otonycteris hemprichii*) as a predator of scorpions: whispering echolocation, passive gleaning and prey selection. *Journal of Comparative Physiology A* 197:425–433.

Howard RJ, Edgecombe GD, Legg DA, Pisani D, and Lozano-Fernandez J. 2019. Exploring the evolution and terrestrialization of scorpions (Arachnida: Scorpiones) with rocks and clocks. *Organisms Diversity Evolution* 19:71–86.

Kerényi C. 1974. Stories of orion. In *The Gods of the Greens*. London: Thames and Hudson.

Klußmann-Fricke B-J, Prendini L, and Wirkner CS. 2012. Evolutionary morphology of the hemolymph vascular system in scorpions: a character analysis. *Arthropod Structure Development* 41 (6):545–560.

Knowlton ED, and Gaffin DD. 2011. Functionally redundant peg sensilla on the scorpion pecten. *Journal of Comparative Physiology A* 197:895–902.

Koch CL. 1837. *Übersicht des arachnidensystems*. Vol. 5. CH Zeh.

Lautie N, Soranzo L, Lajarille M-C, and Stockmann R. 2008. Paraxial organ of a scorpion: structural and ultrastructural studies of *Euscorpius tergestinus* paraxial organ (Scorpiones, Euscorpiidae). *Invertebrate Reproduction Development* 51 (2):77–90.

Lourenço WR. 2008. Parthenogenesis in scorpions: some history-new data. *Journal of Venomous Animals Toxins including Tropical Diseases* 14:19–44.

Lourenço WR. 2018. The evolution and distribution of noxious species of scorpions (Arachnida: Scorpiones). *Journal of Venomous Animals Toxins Including Tropical Diseases* 24:1.

Lourenço WR. 2020. The coevolution between telson morphology and venom glands in scorpions (Arachnida). *Journal of Venomous Animals Toxins Including Tropical Diseases* 26:e20200128.

Lourenço WR, Cloudsley-Thompson JL, Cuellar O, Von Eickstedt VRD, Barraviera B, and Knox MB. 1996. The evolution of scorpionism in Brazil in recent years. *Journal of Venomous Animals Toxins* 2:121–134.

Mattoni CI, Garcia-Hernandez S, Botero-Trujillo R, Ochoa JA, Ojanguren-Affilastro AA, Pinto-da-Rocha R, and Prendini L. 2015. Scorpion sheds 'tail'to escape: consequences and implications of autotomy in scorpions (Buthidae: Ananteris). *PLoS One* 10 (1):e0116639.

Monge-Nájera J. 2019. Scorpion body size, litter characteristics, and duration of the life cycle (Scorpiones). *Cuadernos de Investigación UNED* 11 (2):101–104.

Polis GA. 1990. *The Biology of Scorpions*, edited by GA Polis. CABI Digital Library. Stanford, CA: Stanford University Press.

Rubio M. 2000. Commonly Available Scorpions. *Scorpions: Everything About Purchase, Care, Feeding, and Housing*. Hauppauge, NY: Barron's.

Schofield RMS. 2001. Metals in cuticular structures. In Scorpion Biology Research:234–256.

Soleglad ME, and Fet V. 2003. High-level systematics and phylogeny of the extant scorpions (Scorpiones: Orthosterni). *Euscorpius* 2003 (11):1–56.

Soleglad ME, Fet V, and Kovařík F. 2005. The systematic position of the scorpion genera *Heteroscorpion* Birula, 1903 and *Urodacus* Peters, 1861 (Scorpiones: Scorpionoidea). *Euscorpius* 2005 (20):1–37.

Sridhara S, Chakravarthy AK, Kalarani V, and Reddy DC. 2016. Diversity and ecology of scorpions: evolutionary success through venom. In *Arthropod Diversity Conservation in the Tropics Sub-Tropics*, edited by A. K. Chakravarthy, and S. Sridhara. Singapore: Springer.

Stockmann R. 2015. Introduction to scorpion biology and ecology. In *Scorpion Venoms*, edited by P Gopalakrishnakone, L Possani, EF Schwartz, and R de la Vega. Dordrecht: Springer.

Stockmann R, Ythier E, and Fet V. 2010. *Scorpions of the World*, edited by V. Fet. France: NAP Editions: ISBN 978-2-913688-11-7.

Stockwell SA. 1989. *Revision of the Phylogeny and Higher Classification of Scorpions (Chelicerata)*. Berkeley: University of California, Berkeley ProQuest Dissertations Publishing.

Thompson BM. 2018. The grasshopper mouse and bark scorpion: evolutionary biology meets pain modulation and selective receptor inactivation. *Journal of Undergraduate Neuroscience Education* 16 (2):R51.

van der Meijden A, Lobo Coelho P, Sousa P, and Herrel A. 2013. Choose your weapon: defensive behavior is associated with morphology and performance in scorpions. *PLoS One* 8 (11):e78955.

Wanninger A. 2015. *Evolutionary Developmental Biology of Invertebrates 3: Ecdysozoa I: Non-Tetraconata*, edited by A. Wanninger. Vienna: Springer:ISBN 978-3-7091-1986-0.

Warburg MR. 2010. Reproductive system of female scorpion: a partial review. *The Anatomical Record: Advances in Integrative Anatomy and Evolutionary Biology* 293 (10):1738–1754.

Warburg MR. 2012. Pre- and post-parturial aspects of scorpion reproduction: a review. *European Journal of Entomology* 109 (2):139.

Yigit N, and Benli M. 2010. Fine structural analysis of the stinger in venom apparatus of the scorpion *Euscorpius mingrelicus* (Scorpiones: Euscorpiidae). *Journal of Venomous Animals Toxins Including Tropical Diseases* 16:76–86.

3 Scorpion Venom
Origin, Evolution, Composition, and Functions

3.1 INTRODUCTION

Scorpion envenomation in subtropical and tropical regions can result in various clinical symptoms, making it a severe public health concern (Chippaux and Goyffon 2008; Bahloul et al. 2010; Bawaskar 1984). Venom has evolved into a potent cocktail of bioactive compounds and proteins, which may be valuable in the pharmacology industry for developing new drug prototypes. Scorpion venom contains a wide variety of enzymatic proteins, viz. hyaluronidase, metaloprotease, mucoproteinase, phospholipase, and serine protease; scorpion venom peptides are comprised of disulfide and non-disulfide bridged peptides (DBPs and NDBPs), where DBPs are short-chain toxin (10–50 amino acid residues) and long-chain toxin (30–70 amino acid residues). Such toxins have different medical applications, such as antimicrobial, anticancer, antimalarial, immunosuppressive activities, etc., so drugs can be developed against life-threatening diseases. Due to their high specificity and affinity for ion channels have also been used as a valuable medicinal probe for studying ion channels (Aboutorabi et al. 2016). The non-proteinaceous components of scorpion venom are water, mucosa, nucleotides, mucopolysaccharides, lipids, metals, and inorganic compounds (Ahmadi et al. 2020).

Scorpion venom contains abundant low-molecular-mass peptides (de la Vega, Schwartz, and Possani 2010; Cao et al. 2014; Das, Patra, and Mukherjee 2020; Romero-Gutierrez et al. 2017). Most of these toxins are highly toxic, including neurotoxin, cardiotoxin, nephrotoxin, and hemolytic (Tobassum et al. 2020). However, the toxicity of venom depends on a significant amount (relative abundance) of particular venom components (Huang and Jan 2014; Rao et al. 2015; Restano-Cassulini et al. 2017). Because of this variance in toxicity, nerve and muscle tissues' excitable cells cannot perform their routine tasks, which leads to various physiological events linked to disruptions in metabolism and biology (Al-Asmari, Islam, and Al-Zahrani 2016; Andrikopoulos et al. 2011; Aboutorabi et al. 2016).

The likelihood of complications after a scorpion sting increases in children, older people, and individuals with weakened immune systems compared to normal healthy adults (Ortiz et al. 2015). However, many factors influence the severity of stings, including the scorpion's age and diet, its venom dose, its nutritional status, the mean temperature of the area, geographical variance, the victim's weight and age, their

DOI: 10.1201/9781003540816-3

specific sensitivity, and the location of the sting (Tobassum et al. 2020; Tobassum et al. 2018; Ahmadi et al. 2020).

3.2 THE ORIGIN AND DIVERSITY OF SCORPION TOXIN PEPTIDE SCAFFOLDS: EVOLUTION STINGS

The family tree of the scorpion is elucidated using the forms of the toxins present in its venom. The evolutionary history of the scorpion has been revealed by peptides found in its venom. The results might provide a new perspective on many intricate relationships in nature. Although there are more than 2,400 species of scorpions, scientists are not conclusive in determining how they are all related to each other (Howard 2019). Most scorpion species have relatively similar anatomies, making identifying the exact period of evolutionary divides challenging. However, the sequence of specific venom peptides has been discovered to help identify particular branches of the scorpion family tree (Figure 3.1) (Santibáñez-López et al. 2018).

To better understand how scorpion venom toxins originated, the study used a combination of phylogenomics and molecular modeling to identify relationships between different species of scorpions (Santibáñez-López et al. 2018). Scorpions are an emblematic group of arachnids focused on investigations of arthropod territorialization, morphological stasis, and diversification of body plans (Kjellesvig-Waering 1986; Loret et al. 2001; Sharma et al. 2014; Waddington, Rudkin, and Dunlop 2015). It has long been accepted that, in terms of their origin and physical characteristics, scorpions belong to the most primitive and conservative group of arthropods. Because of the variety of their venom, a complex mixture of bioactive substances (including proteins and peptides) released in specialized organs, researchers and ordinary individuals find scorpions extremely fascinating. Venom interferes with the biochemical and physiological functions of the target organisms (King and Hardy 2013; Casewell et al. 2013; Haney et al. 2016).

It is believed that the emergence of ancient housekeeping genes led to the diversification and neofunctionalization (functional divergence of venom poisons), which is a process fueled by selective pressure (Juarez et al. 2008; Fry et al. 2009; Rokyta et al. 2011; Wong and Belov 2012; Haney et al. 2016; Dowell et al. 2016) (Figure 3.2). Although novel peptides frequently retain the same molecular architecture as their ancestor proteins, substantial functional residue alterations, typically in surface-exposed locations, give rise to newly derived biological activities which may be necessary for the survival of scorpions (Fry et al. 2009; Casewell et al. 2013). Nonetheless, two peptides that share a scaffold but have statistically negligible sequence similarities may evolve to have identical fold shapes, making homology inference difficult (Orengo, Jones, and Thornton 1994).

Through the availability of transcriptomes and scorpion genomes, the development of current-generation sequencing technologies has significantly accelerated the identification of venom diversity. According to scientific consensus, differences in prey preference and/or predatory technique are the primary causes of the dynamic molecular evolution that venom-encoding genes go through (Sunagar et al. 2012; Brust et al. 2013; Casewell et al. 2011; Fry et al. 2003; Fry et al. 2008; Kordiš and

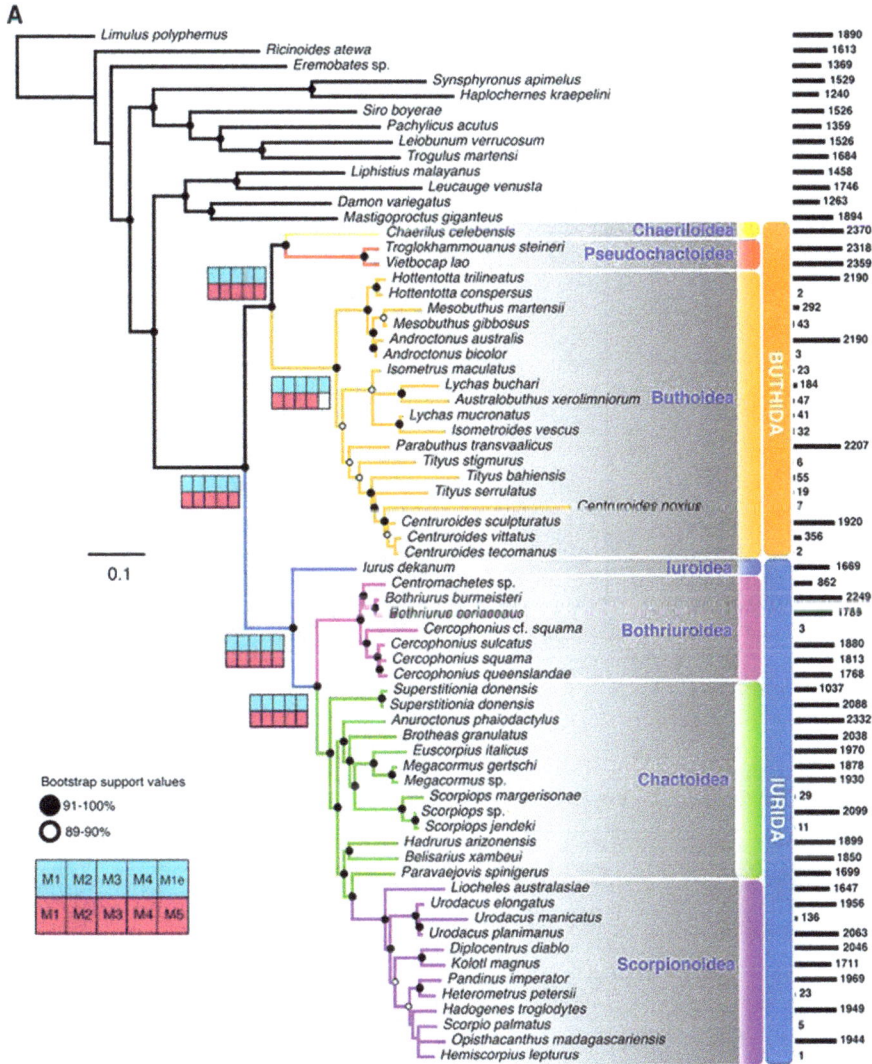

FIGURE 3.1 A diagram showing the life tree of a scorpion. The bars on the right of the terminals indicate the number of orthologs. In Navajo plots, shaded squares denote node recovery from the associated analysis with M = Matrix followed by its number (except M1e = Matrix 1 analyzed with ExaML), and colored as follows: blue squares = IQTree; pink square = ASTRAL. (Figure and legend were adapted with permission from Santibáñez-López et al. 2018; CC BY 4.0.)

Gubenšek 2000; Puillandre, Watkins, and Olivera 2010; Weinberger et al. 2010; Duda and Lee 2009; Barlow et al. 2009; Daltry, Wüster, and Thorpe 1996; Low et al. 2013).

Nevertheless, the underlying genetic diversity of venom may be hidden by the substantial sequence divergence that occurs over extended periods of evolutionary time

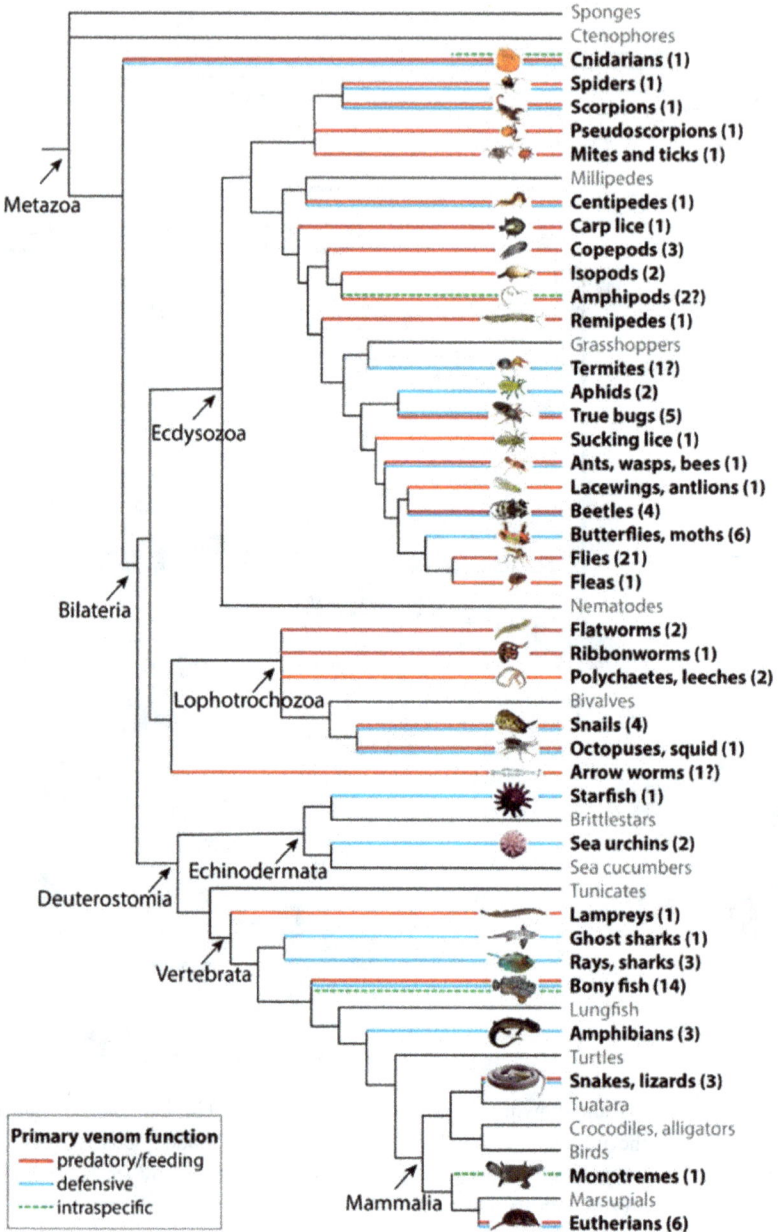

FIGURE 3.2 Taxonomic diversity and the primary functions of scorpion venom. A pruned and schematic phylogenetic tree of venomous animals modified after Casewell et al. (2013), illustrating the frequency. (Figure and legend were adapted with permission from Casewell et al. 2013; License number: 5687091305480.)

and the sporadic pattern of selecting the genes that express it. Some of the earliest noxious lineages, such as coleoids, centipedes, cnidarians, scorpions, spiders, etc., have a high frequency of toxin-encoding genes. Furthermore, the limited taxon selection in biodiscovery-oriented research hinders the full recognition of natural complexity, with easily obtained or medically relevant species often receiving excessive research compared to actual taxonomical diversity.

The venoms that scorpions have developed throughout their more than 400 million years of evolutionary history (Dunlop and Selden 2009) have harmful effects on various biological targets. Because of their long geological history, sluggish migratory rates, and diverse population structures, the toxin diversity has been amplified (Bryson Jr et al. 2013; Yamashita and Rhoads 2013; Zeh 1990; Gantenbein and Largiadèr 2003; Polis 1990). While some snake venoms contain high concentrations of enzymatic toxins, scorpion venoms primarily contain peptide (non-enzymatic) toxins. The cysteine-stabilized (CS)/scaffold is one of several toxin types found in scorpion venoms and is highly complicated. Scorpion CS/toxins include sodium ion channel (NaV) modulators (NaScTx or NaV-CS/toxin; subtypes: -toxins (site-3 binding) and -toxins (site-4 binding)), atypical NaV-CS/toxins (birtoxin and similar toxins, including the alleged "lipolytic toxins"), potassium ion channel (Kv) targeting toxins (KTx).

3.3   STRATEGIES FOR INVESTIGATING VENOM PROTEOMES

Understanding the complex cocktail of scorpion venoms and developing effective antivenom to restore the pathophysiological circumstances caused by pharmacologically active venom components in the victim is the main fascinating issue in the treatment of scorpion envenoming. The absence of expertise and technology for venom decomplexation and the determination of the non-enzymatic components of the venom pose significant, hitherto insurmountable challenges. Nevertheless, the introduction of mass spectrometry (MS) to characterize scorpion venoms at the beginning of the 21st century opened up new avenues for addressing the fundamental issues associated with the complexity of venoms of various species. Similarly, the proteome characterization of various scorpion venoms and antivenom characterization were also investigated by different scientists (Das, Patra, and Mukherjee 2020; Xu et al. 2014; Batista et al. 2006; Romero-Gutiérrez et al. 2018).

Separating venom and antivenom proteins, in general, involves using several protein decomplexation strategies, such as sulfate–polyacrylamide gel electrophoresis (SDS–PAGE) separation and liquid chromatographic separation methods (such as gel-filtration chromatography, ion-exchange chromatography, and Reversed-Phase High Performance Liquid Chromatography or RP-HPLC). Subsequently, for their identification, MS analysis was followed (Das, Patra, and Mukherjee 2020; Das et al. 2022; Batista et al. 2006; Romero-Gutiérrez et al. 2018; Xu et al. 2014).

3.4   QUANTIFICATION OF VENOM TOXINS

3.4.1   LABEL-FREE QUANTIFICATION APPROACH

To date, quantitative determination of protein abundances could be challenging. In contrast, the proteomic shotgun technique for understanding proteomes has led to the comprehensive cataloging of snake and scorpion venom components. Significant advances in non-gel-based quantitative proteomics techniques and the inclusion of isotope-labeled venom peptides as internal standards have made it more accessible and accurate.

Nevertheless, the labeling-based shotgun strategy for toxin quantification has significant drawbacks, such as longer sample preparation times and greater complexity, higher sample concentrations, high reagent costs, insufficient labeling, and the need for specific quantification software (Zhu, Smith, and Huang 2010). Thus, increased focus has shifted toward label-free shotgun proteomic quantification strategies to address these drawbacks and yield faster, cleaner, and more detailed quantification results. Two measurements—peptide peak area or intensity (MS1) and spectral count (MS2)—are used in the label-free method for quantifying identified proteins (Zhu, Smith, and Huang 2010). The peptides are ionized by electrospray ionization (ESI), and ions with a specific mass/charge (m/z) ratio are identified and recorded with a particular intensity and precise retention duration for relative quantification of proteins using the MS1 technique (Zhu, Smith, and Huang 2010). However, this approach has certain practical limitations for complicated biological samples (venoms), such as those with varied peak intensities from run to run and any experimental or chromatographic shift leading to inaccurate quantification.

Several computational methods have been developed to resolve these uncertainties and thoroughly compare the peak intensity data between liquid chromatography-mass spectrometry (LC-MS) runs (Zhu, Smith, and Huang 2010). Due to the simplicity of the MS2-based spectral counting method, in contrast to the chromatographic peak intensity approach, no specific tools or algorithms have been developed; however, for accurate and reliable estimation of relative toxin abundances in venom proteomes, normalization of spectral counts by protein length or the number of identified peptides, as well as proper statistical tests, is required (Zhu, Smith, and Huang 2010; Fox et al. 2006; Shalit et al. 2015). However, because of their relative simplicity of use and minimal sample requirements, MS-based label-free quantification techniques have gained notoriety for recognizing venom toxins (Mukherjee, Kalita, and Mackessy 2016; Tan et al. 2015; Ziganshin et al. 2015; Dutta et al. 2017; Patra et al. 2017). However, these strategies are hindered by database dependencies.

3.4.2   QUANTIFICATION BY THE AREA UNDER RP-HPLC CURVE (AUC)

This method uses SDS-PAGE band intensity and peak areas under the RP-HPLC chromatogram as a surrogate measure to estimate toxin abundances. This method monitors protein elution at 215 nm to assess peptide bond absorption (Lomonte and Calvete 2017; Calvete et al. 2009; Calvete 2014). Comparatively, this method is labor-intensive and slow, mainly when collecting and processing chromatographic fractions. Furthermore, because tyrosine, tryptophan, phenylalanine, and histidine amino acids are present in aromatic side chains, which contribute significantly and may be biased toward the composition of distinct proteins eluted in the RP-HPLC peaks, the side-chain composition of the toxins may also affect the absorbance values at 215 nm (Anthis and Clore 2013). Moreover, due to the limitations of traditional staining, protein components present in trace levels or venom glycoproteins are typically more likely to be overlooked. Furthermore, the high-molecular-weight components can go undetected because they are relatively difficult to elute from the C18 column (Lomonte and Calvete 2017).

3.5 COMPOSITION AND DIVERSE FUNCTIONS OF SCORPION VENOM

Scorpions use their venom to both capture and defend against prey. Scorpion venom is a complex mixture of enzymatic and predominantly non-enzymatic toxins. It also contains various compounds such as water, mucosa, low-molecular-mass peptides, enzymes, free amino acids, biogenic amines, nucleotides, mucopolysaccharides, mucoproteins, histamine, serotonin, heterocyclic components, and several unidentified substances (Tobassum et al. 2018). Both DBPs and NDBPs are present in scorpion venoms, whereas NDBP is a significant component of it (Ortiz et al. 2015).

Based on their chain length, the DBPs are classified into two broad categories: (i) short toxins comprised of 30–40 amino acids constrained by three or four disulfide bridges that block the K^+ channels (Das, Patra, and Mukherjee 2020). These toxins are specific and potent blockers of K^+ channels; that is, charybdotoxin isolated from the scorpion venom is a specific blocker of two isoforms (KCa3.1 and Kv1.3) of potassium (K^+) channel, which is a target for the immunosuppression (Chandy et al. 2004; Chen and Chung 2015) and (ii) long toxins comprised of 60–70 amino acids, cross-linked by four disulfide bridges that affect specifically Na^+ channels, which are a target for the treatment of pain (Knapp, McArthur, and Adams 2012; Chen and Chung 2015; Srairi-Abid et al. 2019; Fontecilla-Camps, Habersetzer-Rochat, and Rochat 1988; Prendini 2000). These peptides are used to develop different drugs (Ortiz et al. 2015). Because of their high affinity and specificity, these ion channel toxins have been used as useful pharmacological probes to study the ion channels (Aboutorabi et al. 2016). Despite their different primary structures, most scorpion toxins have an identical Csαβ (cysteine-stabilized α/β motif) fold (Fontecilla-Camps, Habersetzer-Rochat, and Rochat 1988).

The non-enzymatic toxins found in scorpion venom can be categorized into four types based on their biological function and physiological activity: (i) Na^+ channel toxin, (ii) K^+ channel toxin, (iii) Ca^{2+} channel toxin, and (iv) Cl^- channel toxin (Possani et al. 2000). Specific ions can pass throughout the cell membrane through voltage-gated ion channels (Chen and Chung 2015). They are connected to several physiological functions, including muscle and neuron cell signals. Ion channels are crucial targets for the development of drugs as their dysfunction results in a variety of human disorders (Bagal et al. 2013). The primary mechanism of action of the neurotoxic peptide and protein toxins is to disrupt the normal functioning of ion channels in neuronal and muscle cells, resulting in various clinical symptoms (Bawaskar and Bawaskar 1992; Santos et al. 2016; Rao et al. 2015).

3.6   TOXICITY OF SCORPION VENOM

The signs and symptoms presented by scorpion sting victims result from the complex nature of venom's toxins, which exert hyperactivity to the autonomic nervous system and cause massive damage to vertebrate and invertebrate nervous systems (Gwee et al. 2002).

Victims may exhibit local symptoms of neurotoxic and cytotoxic effects, respectively. The symptoms might also be neurological and non-neurological. Non-neurological symptoms include cardiovascular, pulmonary, gastrointestinal, genitourinary, hemato-logical, and metabolic indications. Most neurological symptoms are caused by either the release of acetylcholine from postganglionic parasympathetic neurons or catechol-amine from the adrenal glands (Freire-Maia, Pinto, and Franco 1974).

Scorpion venom is extremely dangerous because it contains substances that influence ion channels, enzymes, and allergic enzymes, as well as neurotoxins, nephrotoxins, cardiotoxins, and hemolytic toxins (Almeida et al. 2012; Tobassum et al. 2020). Scorpion venom's toxicity is determined by neurotoxins, specifically low-molecular-mass peptides that bind with ion channels (Huang and Jan 2014; Rao et al. 2015; Restano-Cassulini et al. 2017). Typically, they generate the envenoming indications by impairing the ability of excitable cells in muscle and nerve tissues to function normally (Andrikopoulos et al. 2011; Ding et al. 2014; Aboutorabi et al. 2016; Restano-Cassulini et al. 2017). Excitable cells can block the target ion channels (Restano-Cassulini et al. 2017; Ortiz et al. 2015). Additionally, these toxins induce some physiological processes linked to disruptions in metabolism and biology (Al-Asmari, Islam, and Al-Zahrani 2016).

3.7   THE SIGNALS AND THE MARKERS OF THE SCORPION TOXINS DURING ENVENOMATION

The signals of the scorpion toxins are defined via (i) scorpion species, (ii) poison formation, and (iii) the prey's physiological response to the venom. The signs of the bite begin promptly few minutes after the sting and commonly advance to an extreme seriousness within 5 hours. At this duration, the massive liberation of neurosenders results in perspiration, queasiness, and spewing (Mebs 2002). The preys commonly have the main signals, with the most famous being dilated pupils, nystagmus, hype slobber, dysphagia, and insomnia. They may display signs and symptoms, including excitation of the central nervous system, excitation of the autonomic nervous system, respiratory and heart inability, and even death. Afterward, bites via serious scorpions from various portions of the global signals and markers are analogous (Chidambara Murthy, Jayaprakasha, and Singh 2002).

The prey of scorpion intoxicates that offered multiple device-organ inabilities distinguished via alterations in hormonal milieu through a considerable liberation of contrary organizational hormones, for instance, catecholamine, glucagon, cortisol, angiotensin-II, and through reduction grades of insulin and a raised blood glucose level. Estimating the scorpion's intoxication relies on topical signals and whether or not nervous system signals are dominant. The topical symptoms observed in hunting ability disintegrate that ability into anxiolytic and cytotoxic topicals. The central nervous system signals are sympathetic, parasympathetic, physiological, scalp, and marginal nervous systems. The adrenal glands' release of catecholamines or the postganglionic parasympathetic excitable cells' release of acetylcholine is responsible for the highest number of indicators in the nervous system signals (Freire-Maia, Pinto, and Franco 1974).

3.8 VENOM PROTEOME ANALYSES OF SOME VITAL SCORPIONS AROUND THE WORLD

Over the last 10 years, there has been a tangible growth in the scientific methods for identifying, describing, and measuring venom composition in animals. Conventional techniques rely on venom profiling using gel-filtration chromatography, SDS–PAGE, and biochemical studies of venom enzymes. More recently, though, these methods have been coupled with high-throughput genomic, transcriptomic, and proteomics analyses to offer a more complete and in-depth examination of a species' venom (Gutiérrez et al. 1995; Calvete et al. 2009; de la Vega, Schwartz, and Possani 2010; Abdel-Rahman, Quintero-Hernandez, and Possani 2013; Mukherjee, Kalita, and Mackessy 2016; Santibáñez-López et al. 2016; Kalita, Mackessy, and Mukherjee 2018; Saviola, Negrão, and Yates III 2020). The pathophysiology of venomous bites or stings and the content and toxicity of venom are positively correlated, as shown in many studies (Zelanis and Tashima 2014; Chanda and Mukherjee 2020; Manuwar et al. 2020).

Recently, biochemical and in vitro pharmacological activity tests have been used in conjunction with laboratory-based liquid chromatography-mass spectrometry (LC-MS/MS)-based proteomics to characterize the composition of the venom of the Indian red scorpion (*Mesobuthus tamulus*) (Das, Patra, and Mukherjee 2020). Proteomic analysis identified 110 proteins and polypeptides under 13 protein families. The venom contains abundant ion channel toxins (Na^+ and K^+ channels targeting toxins). Other minor venom components are serine protease inhibitors, serine protease-like proteins, hyaluronidase, antimicrobial peptide (AMP), lipolysis potentiating peptides, makatoxin, parabutoporin, neurotoxin affecting Cl^- channels, bradykinin potentiating peptides Ca^{2+} channel toxins, HMG CoA (Hydroxymethylglutaryl-CoA) reductase inhibitor, and several other unknown toxins (Das, Patra, and Mukherjee 2020). The insect-selective low-molecular-mass toxins BtTx3 (3,796 Da) and ButaIT (3,856.7 Da) were also discovered. These toxins have insecticidal properties when applied to lepidopteran insect species (Wudayagiri et al. 2001; Dhawan et al. 2002).

Assessment of pharmacological activity reveals that the Indian red scorpion venom is devoid of in vitro hemolytic activity. It did not demonstrate tested enzyme activity such as phospholipase A_2, L-amino acid oxidase, adenosine tri-, di-, monophosphatase, hyaluronidase, and fibrinogenolytic metalloproteinase. Furthermore, in vitro studies revealed that it does not affect platelet modulation (activation or deaggregation) or blood coagulation. Table 3.1 lists many other venom toxins from various *Mesobuthus* and *Heterometrus* species that are widespread across the Indian subcontinent.

TABLE 3.1
Comparative list of the distribution of venom toxins from different species of *Mesobuthus* and *Heterometrus* throughout the Indian subcontinent

Toxins	*Mesobuthus tamulus*	*Mesobuthus martensii*	*Heterometrus longimanus*	References
Na⁺ channel toxin	YES	YES	YES	(Das, Patra, and Mukherjee 2020; Xu et al. 2014; Bringans et al. 2008)
K⁺ channel toxin	YES	YES	YES	(Das, Patra, and Mukherjee 2020; Xu et al. 2014; Bringans et al. 2008)
Cl⁻ channel toxin	YES	YES	Not known	(Das, Patra, and Mukherjee 2020; Xu et al. 2014)
Ca²⁺ channel toxin	YES	YES	YES	(Xu et al. 2014; Das, Patra, and Mukherjee 2020; Bringans et al. 2008)
Hyaluronidase	YES	YES	Not known	(Das, Patra, and Mukherjee 2020; Xu et al. 2014)
Bukatoxin	YES	Not known	Not known	(Das, Patra, and Mukherjee 2020)
Makatoxin	YES	YES	Not known	(Das, Patra, and Mukherjee 2020; Xu et al. 2014)
Serine protease like protein	YES	YES	Not known	(Xu et al. 2014; Das, Patra, and Mukherjee 2020)
Serine protease inhibitor	YES	YES	Not known	(Xu et al. 2014; Das, Patra, and Mukherjee 2020)
Antimicrobial peptide	YES	YES	YES	(Xu et al. 2014; Das, Patra, and Mukherjee 2020; Bringans et al. 2008)
Lipolysis potentiating peptide	YES	YES	Not known	(Xu et al. 2014; Das, Patra, and Mukherjee 2020)
Parabutoporin	YES	Not known	Not known	(Das, Patra, and Mukherjee 2020)
Bradykinin potentiating peptide	YES	Not known	Not known	(Das, Patra, and Mukherjee 2020)
HMG CoA reductase inhibitor	YES	Not known	Not known	(Das, Patra, and Mukherjee 2020)
Insect toxin	YES	Not known	YES	(Das, Patra, and Mukherjee 2020; Joseph and George 2012)
Dermonecrotic toxin	Not known	Not known	YES	(Bringans et al. 2008)

The Chinese scorpion (*Mesobuthus martensii*) is one of the most populous scorpions in eastern Asian countries. The venom proteome of this species was characterized by 2DE, SDS-PAGE, and HPLC applications where 134 proteins were identified that comprised 43 typical toxins and 7 atypical toxins (including three Na^+ channel toxins, three K^+ channel toxins, and one no-annotation toxin), 72 cell-associated proteins ,and 12 venom enzymes. The molecular weights of the most prevalent proteins were smaller, at approximately 10 kDa. From the crude venom of *M. martensii*, three novel Na^+ channel toxin sequences were identified: comp201_c0_seq1_3, comp162_c0_seq1_6, and MMa37864. Upon examining public databases, the maximum identity of comp201_c0_seq1_3 511 was only 29.58% with neurotoxic 8 (Amm VIII), a long-chain (4 C-C) α-Na^+ channel toxin derived from the venom of the scorpion *Androctonus mauretanicus* (Alami et al. 2006; Alami et al. 2003). The mature peptide Comp162_c0_seq1_6 sequence demonstrated the highest identity of 32.35% with the β-Na^+ channel toxin acra3, which was extracted from the venom of scorpion *Androctonus crassicauda* (Caliskan et al. 2012). MMa37864 had 30.43% identity with β-Na^+ channel toxin Tx273 predicted by scorpion *Buthus occitanus israelis*. These three peptides did not resemble the other known scorpion toxins, indicating that they might be categorized as novel Na^+ channel toxins (Xu et al. 2014).

Proteomic analysis of *Androctonus bicolour* by LC–MS/MS analysis revealed 16 venom peptides, including ion channel toxins and some AMPs (Zhang et al. 2015). One of the deadliest scorpions is *B. occitanus*. By enabling a global perspective of the structural elements of such complex matrices, top-down and bottom-up proteomic analyses are implemented to facilitate screening. A summary of the *B. occitanus* scorpion's venom has been provided below to explore the variety of its toxins and ultimately understand their effects. The nano-high liquid chromatography coupled with nano-electrospray tandem mass spectrometry (nano-LC-ESI MS/MS) was used with top-down and bottom-up strategies.

The LC-MS analysis demonstrated that *B. occitanus* venom contains 200 toxins whose molecular mass ranges from 1,868 to 16,720 Da; among them, the most representative venom peptides were between 5,000 and 8,000 Da. Interestingly, combined top-down and bottom-up LC-MS/MS results showed the finding of several toxins, preferably ion channel toxins, which target the ion channels, including sodium (NaScTxs), potassium (KScTxs), calcium channels (CaScTx), and chloride (ClScTxs), amphipathic peptides, AMPs, hypothetical secreted proteins, and myotropic neuropeptides. This investigation reveals the molecular diversity of *B. occitanus* scorpion venom and determines components that could be pharmacologically active.

The Colombian scorpion *Tityus pachyurus* is toxic to humans and can produce fatal accidents. At least 104 distinct components were identified by MS analysis of *T. pachyurus*. A strong Shaker B (K^+ type channel) blocker was discovered during electrophysiological experiments using heterologously produced ion channels in insect Sf9 cultured cells. It is the third member of subfamily 13, and -KTx13.3 has been suggested as its systematic name. The peptide has the descriptive name Tpa1 and has a molecular mass of 2457 atomic mass units. It has 23 amino acid residues tightly packed together by three disulfide bridges. The mice assay showed clearly the presence of toxic effects of venom peptides. One is Tpa2, which is stabilized by four disulfide bridges and contains 65 amino acid residues and a molecular mass of 7,522.5 atomic mass units. Similar to the beta scorpion toxins, it was found to alter the Na^+-currents of the F-11 and TE671 cells in culture. These findings show that toxic peptides are present in *T. pachyurus* venom and support the notion that encounters with this scorpion species pose a significant risk to people in Colombia (Barona et al. 2006).

3.9   PHARMACOLOGY OF SCORPION VENOM TOXINS

Scorpion poisons probably increase the release of neurotransmitters by blocking K^+ channels or delaying the inactivation of Na^+ channels (Rowan et al. 1992; Narahashi et al. 1972; Vatanpour, Rowan, and Harvey 1993). Na^+ channel toxins can disrupt nervous system function by varying the activation of Na^+ channels in nerve cells. They are categorized as α- and β-neurotoxins (Stevens, Peigneur, and Tytgat 2011). The α-toxin attaches itself to the Na^+ channel's site 3 receptors, preventing inactivation and extending the action potential (Catterall 1976; Catterall 1986; Couraud et al. 1982). Conversely, the β-toxin attaches itself to the Na^+ channel's site 4 and causes the voltage activation to move toward greater negative potentials, which causes the channel to fire repeatedly and spontaneously (Couraud et al. 1982; England and de Groot 2009). Two key constituents of scorpion venom are NDBPs and DBPs, which display a range of pharmacological consequences (Ortiz et al. 2015; Zeng et al. 2004; Zeng, Corzo, and Hahin 2005).

Disulfide-bridged neurotoxic peptides' structure, characteristics, and pharmacological actions have been successfully characterized through several advances to date (Ortiz et al. 2015). NDBPs were not well-known until the past 10 years, but due to their outstanding performance, researchers now pay enough attention to NDPBs (Ortiz et al. 2015; Zeng et al. 2004; Zeng, Corzo, and Hahin 2005). These peptides have strong antibacterial properties and are flexible, often cationic, helical, and amphipathic (Machado et al. 2016).

The 3,459.1 Da toxin tamapin, which was isolated from the venom of an Indian red scorpion, explicitly inhibits the central nervous system's small conductance Ca^{2+}-activated K^+ (SK) channels. However, studies have shown that potency and pharmacological properties are determined by the amidated C-terminal tyrosine residue of tamapin (Pedarzani et al. 2002). Another Indian red scorpion venom peptide, i.e., *Tamulus* toxin, displays a slow K^+ channel time-dependent inactivation that functions on repetitive firing or the prolonged depolarization of action potentials (Strong et al. 2001). Scorpion toxins' Na^+ and K^+ channels primarily mediate the combinatorial actions that cause the autonomic nerves to depolarize intensely and persistently. Consequently, the "autonomic storm" action is brought on by increasing the discharge of autonomic neurotransmitters (Gwee et al. 2002). The intracellular calcium levels drop by the action of the Ca^{2+} channel scorpion toxins. It prevents the smooth muscle cells in the lungs from contracting, resulting in pulmonary hypertension (Fan, Chen, and Liu 2015; Touyz et al. 2018). Moreover, hyaluronidase is a nontoxic enzyme. It enhances the venom's diffusion rate into the tissue of scorpion sting victims, enhancing the local systemic envenomation (Morey, Kiran, and Gadag

2006). Some scorpions, such as *Heterometrus swammerdami/Palamneus gravimanus*, *Hemiscorpius lepturus*, and *Heterometrus fuvipes,* exhibited hyaluronidase activity (Morey, Kiran, and Gadag 2006; Seyedian et al. 2010; Ramanaiah, Parthasarathy, and Venkaiah 1990).

It is also likely that some snake and scorpion venom enzymes have different substrate specificities as they act differently from the conventional serine proteases and hyaluronidases found in snake venom (Thakur and Mukherjee 2015; Cid-Uribe et al. 2020; Das, Patra, and Mukherjee 2020). Moreover, the presence of phospholipase A$_2$ (MtPLA2,19 kDa) in scorpion venom has been reported in some venom of scorpions such as *H. lepturus* (Hemipilin 1 and 2, 15 kDa) (Jridi et al. 2017), *Anuroctonus phaiodactylus* (Phaiodactylipin, 19.1 kDa) (Valdez-Cruz, Batista, and Possani 2004), *Heterometrus laoticus* (HmTx, 14 kDa) (Incamnoi et al. 2013), *Heterometrus fulvipes* (HfPLA2, 16 kDa) (Ramanaiah, Parthasarathy, and Venkaiah 1990), *Scorpio maurus* (phospholipin and Sm-PLGV, 14.8 and 15.15 kDa, respectively) (Conde et al. 1999; Louati et al. 2013; Krayem et al. 2018), and *Pandinus imperator* (IpTxi, 15 kDa) (Zamudio et al. 1997). A wide range of pharmacological properties, including myotoxicity, neurotoxicity, inflammation, hemolysis, as well as anticoagulant, antimicrobial, and antitumor activities, are displayed by PLA$_2$s (Krayem and Gargouri 2020).

3.10 STRUCTURE OF ION CHANNEL AND MECHANISM OF ACTION OF VOLTAGE-GATED ION CHANNEL SCORPION TOXINS

Animals, plants, and bacteria all contain ion channels, which control the flow of ions across cell membranes (Stock, Souza, and Treptow 2013). These channels influence muscle contraction, potential membrane formation, signal transmission, neurotransmitter release, hormone production, chemical and physical stimuli sensing motility, and cell development (Kozlov 2018; Pless and Ahern 2015). They can be divided into groups based on their homologous sequence, ion selectivity, and gating mechanisms for opening and closing. Voltage-gated, ligand-dependent, and mechanically sensitive channels are the three types of gating channels (Zhao et al. 2019).

Voltage-gated Na$^+$, K$^+$, and Ca^{2+} channels have similar structures in mammals and typically comprise a pore-forming subunit (Mendes et al. 2023). The main distinction is that the subunit in Na$^+$ and Ca^{2+} channels comprises four linked domains (DI-IV), whereas the subunit in K$^+$ channels contains the tetramerization of four distinct domains (Luz Morales-Lazaro et al. 2015; Van Theemsche et al. 2020). The component consists of six transmembrane (TM) helix segments (S1–S6) for each domain (Vargas et al. 2012). These domains are further separated into a voltage-sensing domain (VSD) made up of S1–S4 segments, with S4 containing positively charged residues, and a pore-forming domain (PD) made up of S5 and S6 segments (Luz Morales-Lazaro et al. 2015; Van Theemsche et al. 2020). Four VSDs are located throughout each of the four PDs clustered together in the pore. The intracellular activation gate, found at the intersection of the four S6 helices, prevents the entrance

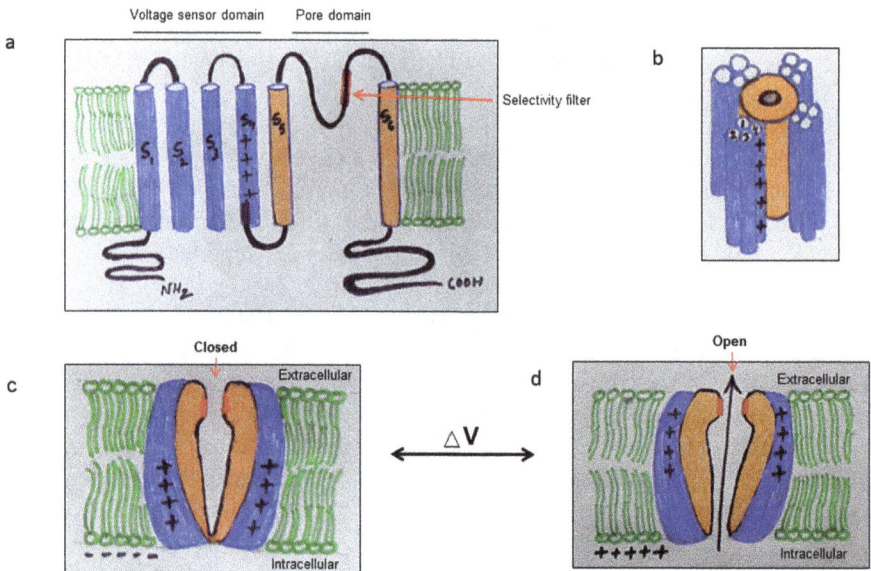

FIGURE 3.3   General architecture of a voltage-gated ion channel. (a) Each subunit comprises six transmembrane helices (S1–S6 lined with intracellular N and C termini). (b) The four subunits tetramerize to shape an ion channel with a pore-forming central unit (orange) enclosed by four VSDs (blue). (c) S4 charges move outward with an alteration in membrane voltage that leads to the opening of the ion channel. (Figure and legend were adapted from Börjesson and Elinder, 2008; sketch by B. Das.)

of ions when it is closed or deactivated and is opened and closed by the VSD (Luz Morales-Lazaro et al. 2015; Van Theemsche et al. 2020) (Figure 3.3).

After a change in the membrane potential, the S4 segment moves outward or inward, causing conformational changes that, in turn, cause the channel pore to open or close as appropriate (Catterall and Swanson 2015). The selectivity filter, which is located at the pore-lining loops (P-loops) connecting S5–S6 in the domains and is made up of conserved residues specific to each channel, enables selectivity for particular ions (Luz Morales-Lazaro et al. 2015). Peptide toxins, which regulate voltage-gated Na^+/K^+ channel activity, are prevalent in the venom of scorpions. Selectivity for sodium in channels is determined by four amino acid residues (DEKA), which are present in an analogous location in each of the domains (Catterall 2017). In potassium channels, the selectivity filter consists of the conserved signature sequence TVGYG in the P-loop (Choe 2002; Kuang, Purhonen, and Hebert 2015).

Several different voltage-gated ion channel types are involved in the action potential the electrical signal nerve cells produce (Bean 2007). Our understanding of the action potential is based on the analysis of the squid axon (Hodgkin and Huxley 1952), in which voltage-gated Na^+ (Nav) channels open for a short period before rapidly deactivating; after a short while, Kv channels are activated and remain open for a longer interval (Armstrong and Hille 1998; Bean 2007). Na^+ ions flow into the cell, and K^+ ions flow into the extracellular environment due to a traveling action potential upon depolarization. Action potentials serve different purposes in the neuron cell bodies and axons, and various types of neurons also have unique action potential patterns (Bean 2007).

The toxins found in scorpion venom are similar in structure and have comparable physicochemical properties. However, due to spontaneous evolution and natural selection, they possess unique pharmacological properties that have developed over millions of years. They subdue animals with neurotoxins that may block or control ion channels (Srairi-Abid et al. 2019; de Lera Ruiz and Kraus 2015). The primary mechanism causing the pharmacological effects of neurotoxins is their interaction with ion channels (Srairi-Abid et al. 2019). Peptides operating on ion channels may have a crucial and functional role during the evolution of scorpion venom in the Buthidae family of neurotoxins, which are incredibly toxic and vary in their preference for ion channels found in mammals and arthropods, as well as within and between species. This fact is particularly relevant concerning scorpions, which share a common ancestor with them (So et al. 2021).

3.10.1 K^+ Channel Blockers and Mechanism of Action

The four primary categories of potassium channels are voltage-gated, tandem pore domain, inwardly rectifying, and calcium-triggered. Although these channels have similar structures, the distinctions between these types primarily relate to how the gate receives its signal (Kuang, Purhonen, and Hebert 2015). "K^+ channel toxins" are scorpion toxins that inhibit various K^+ channels, and these toxins have been extensively explored (Harvey et al. 1995; Quintero-Hernández et al. 2013). Although shorter than Nav toxins, they share many similarities (Ortiz et al. 2015). K^+ channel-specific toxins (KTxs), exploited in the structural and functional characterization of several K^+ channels, are abundant in scorpion venom. KTx has been divided into four families: α-, β-, γ-, and κ-KTx based on the primary amino acid sequences and cysteine pairing (Tytgat et al. 2000; de la Vega and Possani 2004).

Numerous KTxs from *Mesobuthus eupeus* and *M. martensii* scorpions have been isolated (Shi et al. 2008). These KTxs block various K^+ channels, including neurotoxins, which act on the Kv1.2 channel (Yi et al. 2008), BeKm-1, which blocks the hERG channel (Yi et al. 2007; Yi et al. 2008), charybdotoxin acts on the BKCa channel (Qiu et al. 2009), Kunitz-type toxins block the Kv1.3 channel (Chen et al. 2012), and BmP05, which blocks the SKCa3 channel (Han et al. 2008; Yin et al. 2008). There have also been reports of many active scorpion venom components acting on Cl^- and Ca^{2+} channels (Possani et al. 2000; Guéguinou et al. 2014; Rao et al. 2015).

Pore-blocking peptides form a strong bond with the K^+ channel's outer vestibule, blocking the channel's selectivity filter and preventing K^+ ions from being transported (Chen and Chung 2015; Oukkache et al. 2015). The channel protein's conformational variations determine the mechanism of K^+ channel activity. The ion channel is closed at the resting membrane potential and does not conduct ions. Increases in membrane potential impact the voltage sensor, which opens the channel (Bezanilla 2000). The ion channel responds to even minute changes in membrane potential because of the

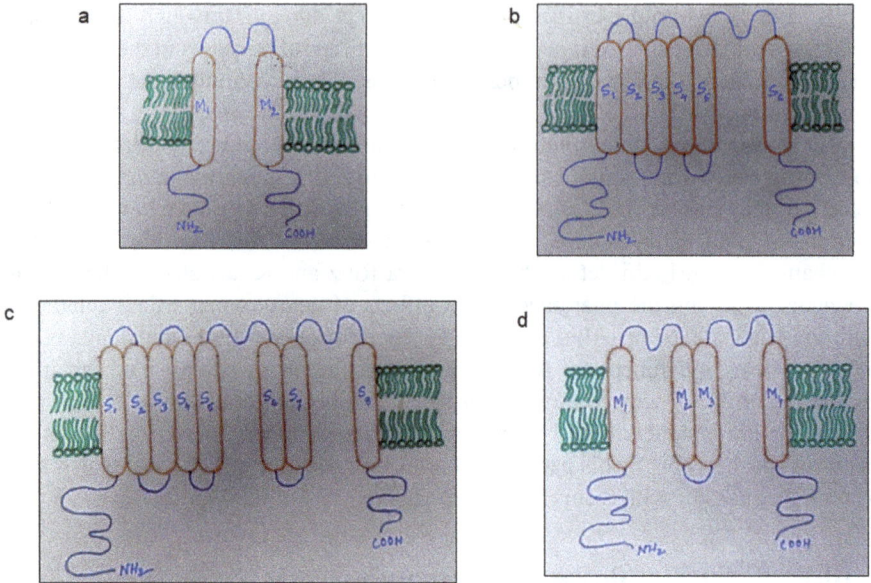

FIGURE 3.4 The four primary classes of K+ channels. (a) 2TM/P channels, (b) 6TM/P channels, (c) 8TM/2P channels, (d) 4TM/2P channels. Positive signs on S4 denote its function in voltage sensing in voltage-gated K+ channels. The channel maintains its open conformational state in the N-type inactivation, but the alpha subunit's N terminal blocks the pore. The N-type mechanism's inactivation is abolished when the N terminal fragment is removed and restored when the fragment is added as a peptide. In the meantime, the C-type of inactivation does not include the N terminal. This kind of inactivation is caused by structural components found in the vestibule of the selectivity filter (Kuzmenkov, Grishin, and Vassilevski 2015). (Figure and legend were adapted from Choe, 2002; sketch by B. Das.)

voltage sensor's high sensitivity, and it conducts ions until it reaches the inactivation phase (Kuzmenkov, Grishin, and Vassilevski 2015). Two processes (N-type and C-type) are involved in the inactivation of the Kv channel (Figure 3.4).

3.10.2 NA+ CHANNEL BLOCKERS AND MECHANISM OF NEUROTOXIN BINDING WITH NA+ CHANNEL

The rapid inflow of Na+ ions increases the action potential in muscle, nerve, and endocrine cells caused by voltage-gated sodium channels (VGSCs) (Catterall et al. 2007). The body's excitable cells (muscles, neurons, and endocrine cells) are distributed with VGSC isoforms, each associated with distinct characteristics in the relevant cells and tissues. In contrast, Nav1.9, 1.8, and 1.7 isoforms are expressed in the peripheral nervous system, and lastly, Nav1.5 and 1.4 are substantially represented in the heart and skeletal muscles, respectively. The central nervous system contains Nav1.6, 1.3, 1.2, and 1.1 isoforms (Stevens, Peigneur, and Tytgat 2011). The term "Nav channel long-chain toxin" refers to highly lethal toxins that alter the voltage-gated Na+ channel

(Nav) (de la Vega and Possani 2005; Andrikopoulos et al. 2011; Davis et al. 2012; Aboutorabi et al. 2016).

According to de la Vega and Possani (2005) and Cao et al. (2014), Nav channel long-chain toxins modulate the activation and inactivation of sodium channels by - NaTxs and -NaTxs, respectively. Sodium toxin (BmKIM) obtained from scorpion *M. martensii* inhibited sodium currents in rat ganglion neurons and myocytes and prevented cardiac arrhythmia in a mouse model (Peng et al. 2002).

Sodium channels, transmembrane complexes, consist of two subunits: the large core protein is α-subunit (220–260 kDa), associated with a different small regulatory unit, i.e., β-subunit (22–36 kDa). The pore containing α-subunit is selectively permeable to Na^+ ions, composed of four homologous domains (DI–DIV). These include six transmembrane segments (S1–S6) (Yu and Catterall 2003). Three cytoplasmic loops connect these four domains to create a bell-shaped protein (Figure 3.5). The voltage-sensing module generated by S1–4 is the first module found in each of the four domains (DI–DIV). The second is the pore-producing module developed by S5 and S6 and the connecting loop (Sato et al. 2001). There are two ways in which toxins affect VGSCs. When the neurotoxin physically blocks the pore and reduces sodium ion conductance, it either causes a blockage of the pore or a modification of the gating that changes the voltage dependence and gating kinetics of the ion channels. The first method is used by toxins when they interact with site 1. For instance, site 1 pore blockers include tetrodotoxin (TTX) and sexitoxin (STX) (Tobassum et al. 2018).

Site 2 toxins such as grayanotoxin and batrachotoxin block inactivation, causing channels to stay continuously active (Stevens, Peigneur, and Tytgat 2011). Toxins from sea anemones and scorpions bind to site 3 and prevent the inactivation (Possani

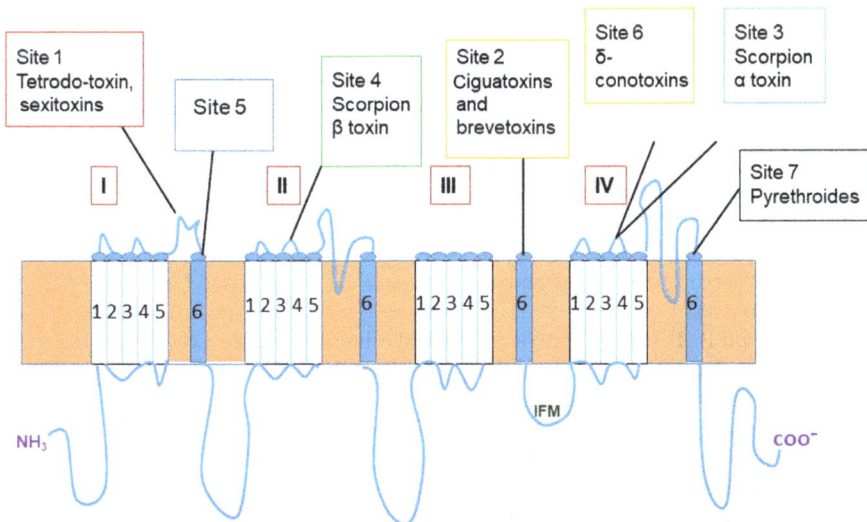

FIGURE 3.5 Identification of the neurotoxic binding regions and schematic representation of the voltage-gated sodium channel (VGSC) α-subunit. (Figure and legend were adapted from Stevens et al. 2011; sketch by B. Das.)

et al. 2000). Site 4 toxins, like those seen in scorpions and spiders, cause the activation to become hyperpolarized (Shichor et al. 2002). When associating with VGSC, site 5 toxins such as ciguatoxins and brevetoxins show a noticeable effect, such as inhibition of activation and the hyperpolarizing shift of voltage-dependent activation. Finally, through blocking inactivation, δ-conotoxins interact with site 6 and have effects comparable to those of the neurotoxins that affect site 3 (Figure 3.5) (Stevens, Peigneur, and Tytgat 2011). Proteases and protease inhibitors are some other components found in scorpion venom in addition to these neurotoxins.

3.11 CONCLUSION

Scorpion venom is a complex collection of biomolecules that can disrupt the victim's physiological function after envenomation. Ion channels regulate the passage of ions across cell membranes in animals, plants, and microorganisms. These channels facilitate potential membrane formation, signal transmission, neurotransmitter release, hormone production, muscle contraction, chemical and physical stimuli perception, motility, and cell development. Scorpion venoms are mainly neurotoxic: nephrotoxin, cardiotoxin, and hemolytic, which target primarily voltage-gated Na^+/K^+ channels that influence their function. The complex nature of the various scorpion venom components causes other signs and symptoms, leading to different clinical manifestations in the victim's body.

REFERENCES

Abdel-Rahman MA, Quintero-Hernandez V, and Possani LD. 2013. Venom proteomic and venomous glands transcriptomic analysis of the Egyptian scorpion *Scorpio maurus palmatus* (Arachnida: Scorpionidae). *Toxicon* 74:193–207.

Aboutorabi A, Naderi N, Vatanpour H, and Zolfagharian H. 2016. Voltage-gated sodium channels modulation by *Bothutous schach* scorpion venom. *Iranian Journal of Pharmaceutical Sciences* 12 (3):55–64.

Ahmadi S, Knerr JM, Argemi L, Bordon KCF, Pucca MB, Cerni FA, Arantes EC, Çalışkan F, and Laustsen AH. 2020. Scorpion venom: detriments and benefits. *Biomedicines* 8 (5):118.

Al-Asmari AK, Islam M, and Al-Zahrani. 2016. In vitro analysis of the anticancer properties of scorpion venom in colourectal and breast cancer cell lines. *Oncology Letters* 11 (2):1256–1262.

Alami M, Ceard B, Legros C, Bougis PE, and Martin-Euaclaire MF. 2006. Genomic characterisation of the toxin Amm VIII from the scorpion *Androctonus mauretanicus mauretanicus*. *Toxicon* 47 (5):531–536.

Alami M, Vacher H, Bosmans F, Devaux C, Rosso JP, Bougis PE, Tytgat J, Darbon H, and Martin-Euaclaire MF. 2003. Characterization of Amm VIII from *Androctonus mauretanicus mauretanicus*: a new scorpion toxin that discriminates between neuronal and skeletal sodium channels. *Biochemical Journal* 375 (3):551–560.

Almeida DD, Scortecci KC, Kobashi LS, Agnez-Lima LF, Medeiros SR, Silva-Junior AA, Junqueira-de-Azevedo ID, Fernandes-Pedrosa MD. 2012. Profiling the resting venom gland of the scorpion *Tityus stigmurus* through a transcriptomic survey. *BMC Genomics* 13 (1):1–11.

Andrikopoulos P, Fraser SP, Patterson L, Ahmad Z, Burcu H, Ottaviani D, Diss JK, Box C, Eccles SA, and Djamgoz MB. 2011. Angiogenic functions of voltage-gated Na$^+$ channels in human endothelial cells: modulation of vascular endothelial growth factor (VEGF) signaling. *Journal of Biological Chemistry* 286 (19):16846–16860.

Anthis NJ, and Marius Clore G. 2013. Sequence-specific determination of protein and peptide concentrations by absorbance at 205 nm. *Protein Science* 22 (6):851–858.

Armstrong CM, and Hille. 1998. Voltage-gated ion channels and electrical excitability. *Neuron* 20 (3):371–380.

Bagal SK, Brown AD, Cox P.J, Omoto K, Owen RM, Pryde DC, Sidders B, Skerratt SE, Stevens EB, and Storer RI. 2013. Ion channels as therapeutic targets: a drug discovery perspective. *Journal of Medicinal Chemistry* 56:593–624.

Bahloul M, Chabchoub I, Chaari A, Chtara K, Kallel H, Dammak H, Ksibi H, Chelly H, Rekik N, Hamida CB, and Bouaziz M. 2010. Scorpion envenomation among children: clinical manifestations and outcome (analysis of 685 cases). *The American Journal of Tropical Medicine Hygiene* 83 (5):1084.

Barlow A, Pook CE, Harrison RA, and Wüster W. 2009. Coevolution of diet and prey-specific venom activity supports the role of selection in snake venom evolution. *Proceedings of the Royal Society B: Biological Sciences* 276 (1666):2443–2449.

Barona J, Batista CV, Zamudio FZ, Gomez-Lagunas F, Wanke E, Otero R, and Possani LD. 2006. Proteomic analysis of the venom and characterization of toxins specific for Na$^+$- and K$^+$-channels from the Colombian scorpion *Tityus pachyurus*. *Biochimica et Biophysica Acta (BBA)—Proteins and Proteomics* 1764 (1):76–84.

Batista CV, D'Suze G, Gómez-Lagunas F, Zamudio FZ, Encarnación S, Sevcik C, and Possani LD. 2006. Proteomic analysis of *Tityus discrepans* scorpion venom and amino acid sequence of novel toxins. *Proteomics* 6 (12):3718–3727.

Bawaskar HS. 1984. Scorpion sting. *Transactions of the Royal Society of Tropical Medicine Hygiene* 78 (3):414–415.

Bawaskar HS, and Bawaskar PH. 1992. Management of the cardiovascular manifestations of poisoning by the Indian red scorpion (*Mesobuthus tamulus*). *Heart* 68 (11):478–480.

Bean BP. 2007. The action potential in mammalian central neurons. *Nature Reviews Neuroscience* 8 (6):451–465.

Bezanilla F. 2000. The voltage sensor in voltage-dependent ion channels. *Physiological Reviews* 80 (2):555–592.

Börjesson SI, and Elinder F. 2008. Structure, function, and modification of the voltage sensor in voltage-gated ion channels. *Cell Biochemistry and Biophysics* 52:149–174.

Bringans S, Eriksen S, Kendrick T, Gopalakrishnakone P, Livk A, Lock R, and Lipscombe R. 2008. Proteomic analysis of the venom of *Heterometrus longimanus* (Asian black scorpion). *Journal of Proteomics* 8 (5):1081–1096.

Brust A, Sunagar K, Undheim EA, Vetter I, Yang DC, Casewell NR, Jackson TN, Koludarov I, Alewood PF, Hodgson WC, and Lewis RJ. 2013. Differential evolution and neofunctionalization of snake venom metalloprotease domains. *Molecular Cellular Proteomics* 12 (3):651–663.

Bryson Jr RW, Riddle BR, Graham MR, Smith BT, and Prendini L. 2013. As old as the hills: montane scorpions in southwestern North America reveal ancient associations between biotic diversification and landscape history. *PLoS One* 8 (1):e52822.

Caliskan F, García BI, Coronas FI, Restano-Cassulini R, Korkmaz F, Sahin Y, Corzo G, and Possani LD. 2012. Purification and cDNA cloning of a novel neurotoxic peptide (Acra3) from the scorpion *Androctonus crassicauda*. *Peptides* 37 (1):106–112.

Calvete JJ. 2014. The expanding universe of mass analyzer configurations for biological analysis. *Plant Proteomics: Methods Protocols* 1072:61–81.

Calvete JJ, Sanz L, Angulo Y, Lomonte B, and Gutiérrez JM. 2009. Venoms, venomics, antivenomics. *FEBS Letters* 583:1736–1743.

Cao Z, Di Z, Wu Y, and Li W. 2014. Overview of scorpion species from China and their toxins. *Toxins* 6 (3):796–815.

Casewell NR, Wagstaff SC, Harrison RA, Renjifo C, and Wüster W. 2011. Domain loss facilitates accelerated evolution and neofunctionalization of duplicate snake venom metalloproteinase toxin genes. *Molecular Biology Evolution* 28 (9):2637–2649.

Casewell NR, Wüster W, Vonk FJ, Harrison RA, and Fry BG. 2013. Complex cocktails: the evolutionary novelty of venoms. *Trends in Ecology Evolution* 28 (4):219–229.

Catterall WA. 1976. Purification of a toxic protein from scorpion venom which activates the action potential Na$^+$ ionophore. *Journal of Biological Chemistry* 251 (18):5528–5536.

Catterall WA. 1986. Molecular properties of voltage-sensitive sodium channels. *Annual Review of Biochemistry* 55 (1):953–985.

Catterall WA, 2017. Forty years of sodium channels: structure, function, pharmacology, and epilepsy. *Neurochemical Research* 42:2495–2504.

Catterall WA, Cestèle S, Yarov-Yarovoy V, Yu Frank H, Konoki K, and Scheuer T. 2007. Voltage-gated ion channels and gating modifier toxins. *Toxicon* 49 (2):124–141.

Catterall WA, and Teresa M. 2015. Structural basis for pharmacology of voltage-gated sodium and calcium channels. *Molecular Pharmacology* 88 (1):141–150.

Chanda A, and Mukherjee AK. 2020. Mass spectrometric analysis to unravel the venom proteome composition of Indian snakes: opening new avenues in clinical research. *Expert Review of Proteomics* 17 (5):411–423.

Chandy KG, Wulff H, Beeton C, Pennington M, Gutman GA, and Cahalan MD. 2004. K$^+$ channels as targets for specific immunomodulation. *Trends in Pharmacological Sciences* 25 (5):280–289.

Chen R, and Chung SH. 2015. Computational studies of venom peptides targeting potassium channels. *Toxins* 7 (12):5194–5211.

Chen ZY, Hu YT, Yang WS, He YW, Feng J, Wang B, Zhao RM, Ding JP, Cao ZJ, and Li WX. 2012. Hg1, novel peptide inhibitor specific for Kv1. 3 channels from first scorpion Kunitz-type potassium channel toxin family. *Journal of Biological Chemistry* 287:13813–13821.

Chippaux JP, and Goyffon M. 2008. Epidemiology of scorpionism: a global appraisal. *Acta Tropica* 107 (2):71–79.

Choe S, 2002. Potassium channel structures. *Nature Reviews Neuroscience* 3:115–121.

Cid-Uribe JI, Veytia-Bucheli JI, Romero-Gutierrez T, Ortiz E, and Possani LD. 2020. Scorpion venomics: a 2019 overview. *Expert Review of Proteomics* 17 (1):67–83.

Conde R, Zamudio FZ, Becerril B, and Possani LD. 1999. Phospholipin, a novel heterodimeric phospholipase A$_2$ from *Pandinus imperator* scorpion venom. *FEBS Letters* 460 (3):447–450.

Couraud F, Jover E, Dubois JM, and Rochat H. 1982. Two types of scorpion toxin receptor sites, one related to the activation, the other to the inactivation of the action potential sodium channel. *Toxicon* 20 (1):9–16.

Daltry JC, Wüster W, and Thorpe RS. 1996. Diet and snake venom evolution. *Nature* 379 (6565):537–540.

Das B, Patra A, and Mukherjee AK. 2020. Correlation of venom toxinome composition of Indian red scorpion (*Mesobuthus tamulus*) with clinical manifestations of scorpion stings: failure of commercial antivenom to immune-recognize the abundance of low molecular mass toxins of this venom. *Journal of Proteome Research* 19 (4):1847–1856.

Das B, Patra A, Puzari U, Deb P, and Mukherjee AK. 2022. In vitro laboratory analyses of commercial anti-scorpion (*Mesobuthus tamulus*) antivenoms reveal their quality and safety but the prevalence of a low proportion of venom-specific antibodies. *Toxicon* 215:37–48.

Davis GC, Kong Y, Paige M, Li Z, Merrick EC, Hansen T, Suy S, Wang K, Dakshanamurthy S, Cordova A, and McManus OB. 2012. Asymmetric synthesis and evaluation of a hydroxyphenylamide voltage-gated sodium channel blocker in human prostate cancer xenografts. *Bioorganic Medicinal Chemistry* 20 (6):2180–2188.

de la Vega RC, Schwartz EF, and Possani LD. 2010. Mining on scorpion venom biodiversity. *Toxicon* 56:1155–1161.

de la Vega RCR, and Possani LD. 2004. Current views on scorpion toxins specific for K$^+$-channels. *Toxicon* 43 (8):865–875.

de la Vega RCR, and Possani LD. 2005. Overview of scorpion toxins specific for Na$^+$ channels and related peptides: biodiversity, structure–function relationships and evolution. *Toxicon* 46 (8):831–844.

de Lera Ruiz M, and Kraus RL. 2015. Voltage-gated sodium channels: structure, function, pharmacology, and clinical indications. *Journal of Medicinal Chemistry* 58 (18):7093–7118.

Dhawan R, Joseph S, Sethi A, and Lala AK. 2002. Purification and characterization of a short insect toxin from the venom of the scorpion *Buthus tamulus*. *FEBS Letters* 528 (1–3):261–266.

Ding J, Chua P-J, Bay B-H, and Gopalakrishnakone P. 2014. Scorpion venoms as a potential source of novel cancer therapeutic compounds. *Experimental Biology Medicine* 239 (4):387–393.

Dowell NL, Giorgianni MW, Kassner VA, Selegue JE, Sanchez EE, and Carroll SB. 2016. The deep origin and recent loss of venom toxin genes in rattlesnakes. *Current Biology* 26 (18):2434–2445.

Duda TF, and Lee T. 2009. Ecological release and venom evolution of a predatory marine snail at Easter Island. *PLoS One* 4 (5):e5558.

Dunlop JA, and Selden PA. 2009. Calibrating the chelicerate clock: a paleontological reply to Jeyaprakash and Hoy. *Experimental Applied Acarology* 48:183–197.

Dutta S, Chanda A, Kalita B, Islam T, Patra A, and Mukherjee AK. 2017. Proteomic analysis to unravel the complex venom proteome of eastern India *Naja naja*: correlation of venom composition with its biochemical and pharmacological properties. *Journal of Proteomics* 156:29–39.

England S., and de Groot MJ. 2009. Subtype-selective targeting of voltage-gated sodium channels. *British Journal of Pharmacology* 158:1413–1425.

Fan Z, Chen Y, and Liu H. 2015. Calcium channel blockers for pulmonary arterial hypertension. *Cochrane Database of Systematic Reviews* 9:CD010066.

Fontecilla-Camps JC, Habersetzer-Rochat C, and Rochat H. 1988. Orthorhombic crystals and three-dimensional structure of the potent toxin II from the scorpion *Androctonus australis* Hector. *Proceedings of the National Academy of Sciences* 85 (20):7443–7447.

Fox JW, Ma L, Nelson K, Sherman NE, and Serrano SMT. 2006. Comparison of indirect and direct approaches using ion-trap and Fourier transform ion cyclotron resonance mass spectrometry for exploring viperid venom proteomes. *Toxicon* 47 (6):700–714.

Freire-Maia L, Pinto GI, and Franco I. 1974. Mechanism of the cardiovascular effects produced by purified scorpion toxin in the rat. *Journal of Pharmacology Experimental Therapeutics* 188 (1):207–213.

Fry BG, Roelants K, Champagne DE, Scheib H, Tyndall JD, King GF, Nevalainen TJ, Norman JA, Lewis RJ, Norton RS, and Renjifo C. 2009. The toxicogenomic multiverse: convergent recruitment of proteins into animal venoms. *Annual Review of Genomics Human Genetics* 10:483–511.

Fry BG, Scheib H, van der Weerd L, Young B, McNaughtan J, Ramjan SFR, Vidal N, Poelmann RE, and Norman JA. 2008. Evolution of an arsenal. *Molecular and Cellular Proteomics* 7 (2):215–246.

Fry BG, Wüster W, Kini RM, Brusic V, Khan A, Venkataraman D, and Rooney AP. 2003. Molecular evolution and phylogeny of elapid snake venom three-finger toxins. *Journal of Molecular Evolution* 57:110–129.

Gantenbein B, and Largiadèr CR. 2003. The phylogeographic importance of the Strait of Gibraltar as a gene flow barrier in terrestrial arthropods: a case study with the scorpion *Buthus occitanus* as model organism. *Molecular Phylogenetics Evolution* 28 (1):119–130.

Guéguinou M, Chantôme A, Fromont G, Bougnoux P, Vandier C, and Potier-Cartereau M. 2014. KCa and Ca²⁺ channels: the complex thought. *Biochimica et Biophysica Acta—Molecular Cell Research* 1843 (10):2322–2333.

Gutiérrez JM, Romero M, Díaz C, Borkow G, and Ovadia M. 1995. Isolation and characterization of a metalloproteinase with weak hemorrhagic activity from the venom of the snake *Bothrops asper* (terciopelo). *Toxicon* 33 (1):19–29.

Gwee MCE, Nirthanan S, Khoo H-E, Gopalakrishnakone P, Kini RM, and Cheah L-S. 2002. Autonomic effects of some scorpion venoms and toxins. *Clinical Experimental Pharmacology Physiology* 29 (9):795–801.

Han S, Yi H, Yin SJ, Chen ZY, Liu H, Cao ZJ, Wu YL, and Li WX. 2008. Structural basis of a potent peptide inhibitor designed for Kv1. 3 channel, a therapeutic target of autoimmune disease. *Journal of Biological Chemistry* 283:19058–19065.

Haney RA, Clarke TH, Gadgil R, Fitzpatrick R, Hayashi CY, Ayoub NA, and Garb JE. 2016. Effects of gene duplication, positive selection, and shifts in gene expression on the evolution of the venom gland transcriptome in widow spiders. *Genome Biology Evolution* 8 (1):228–242.

Harvey AL, Vatanpour H, Rowan EG, Pinkasfeld S, Vita C, Ménez A, and Martin-Eauclaire M-F. 1995. Structure–activity studies on scorpion toxins that block potassium channels. *Toxicon* 33 (4):425–436.

Hodgkin AL, and Huxley AF. 1952. A quantitative description of membrane current and its application to conduction and excitation in nerve. *Journal of Physiology* 117 (4):500.

Howard RJ, Edgecombe GD, Legg DA, Pisani D, & Lozano-Fernandez J. 2019. Exploring the evolution and terrestrialization of scorpions (Arachnida: Scorpiones) with rocks and clocks. *Organisms Diversity Evolution* 19:71–86.

Huang X, and Jan LY. 2014. Targeting potassium channels in cancer. *Journal of Cell Biology* 206 (2):151–162.

Incamnoi P, Patramanon R, Thammasirirak S, Chaveerach A, Uawonggul N, Sukprasert S, Rungsa P, Daduang J, and Daduang S. 2013. Heteromtoxin (HmTx), a novel heterodimeric phospholipase A₂ from *Heterometrus laoticus* scorpion venom. *Toxicon* 61:62–71.

Joseph B, and George J. 2012. Scorpion toxins and its applications. *International Journal of Toxicological and Pharmacological Research* 4 (3):57–61.

Jridi I, Catacchio I, Majdoub H, Shahbazzadeh D, El Ayeb M, Frassanito MA, Solimando AG, Ribatti D, Vacca A, and Borchani L. 2017. The small subunit of Hemilipin2, a new heterodimeric phospholipase A₂ from *Hemiscorpius lepturus* scorpion venom, mediates the antiangiogenic effect of the whole protein. *Toxicon* 126:38–46.

Juarez P, Comas I, Gonzalez-Candelas F, and Calvete JJ. 2008. Evolution of snake venom disintegrins by positive Darwinian selection. *Molecular Biology Evolution* 25 (11):2391–2407.

Kalita B, Mackessy SP, and Mukherjee AK. 2018. Proteomic analysis reveals geographic variation in venom composition of Russell's Viper in the Indian subcontinent: implications for clinical manifestations post-envenomation and antivenom treatment. *Expert Review of Proteomics* 15 (10):837–849.

King GF, and Hardy MC. 2013. Spider-venom peptides: structure, pharmacology, and potential for control of insect pests. *Annual Review of Entomology* 58:475–496.

Kjellesvig-Waering EN. 1986. *A Restudy of the Fossil Scorpionida of the World*. Vol. 55. Paleontological Research Institution.

Knapp O, McArthur JR, and Adams DJ. 2012. Conotoxins targeting neuronal voltage-gated sodium channel subtypes: potential analgesics? *Toxins* 4 (11):1236–1260.

Kordiš D, and Gubenšek F. 2000. Adaptive evolution of animal toxin multigene families. *Gene* 261 (1):43–52.

Kozlov S. 2018. Animal toxins for channelopathy treatment. *Neuropharmacology* 132:83–97.

Krayem N, and Gargouri YJT. 2020. Scorpion venom phospholipases A_2: a minireview. *Toxicon* 184:48–54.

Krayem N, Parsiegla G, Gaussier H, Louati H, Jallouli R, Mansuelle P, Carrière F, and Gargouri Y. 2018. Functional characterization and FTIR-based 3D modeling of full length and truncated forms of *Scorpio maurus* venom phospholipase A_2. *Biochimica et Biophysica Acta General Subjects* 1862 (5):1247–1261.

Kuang Q, Purhonen P, and Hebert H. 2015. Structure of potassium channels. *Cellular Molecular Life Sciences* 72:3677–3693.

Kuzmenkov AI, Grishin EV, and Vassilevski AA. 2015. Diversity of potassium channel ligands: focus on scorpion toxins. *Biochemistry* 80:1764–1799.

Lomonte B, and Calvete JJ. 2017. Strategies in 'snake venomics' aiming at an integrative view of compositional, functional, and immunological characteristics of venoms. *Journal of Venomous Animals Toxins Including Tropical Diseases* 23:26.

Loret E, Hammock BD, Brownell P, and Polis GA. 2001. *Scorpion Biology and Research*, edited by B Philip and P Gary. New York: Oxford University Press, Inc.

Louati H, Krayem N, Fendri A, Aissa I, Sellami M, Bezzine S, and Gargouri Y. 2013. A thermoactive secreted phospholipase A_2 purified from the venom glands of *Scorpio maurus*: relation between the kinetic properties and the hemolytic activity. *Toxicon* 72:133–142.

Low DH, Sunagar K, Undheim EA, Ali SA, Alagon AC, Ruder T, Jackson TN, Gonzalez SP, King GF, Jones A, and Antunes A. 2013. Dracula's children: molecular evolution of vampire bat venom. *Journal of Proteomics* 89:95–111.

Luz Morales-Lazaro S, Hernández-García E, Serrano-Flores B, and Rosenbaum T. 2015. Organic toxins as tools to understand ion channel mechanisms and structure. *Current Topics in Medicinal Chemistry* 15 (7):581–603.

Machado RJ, Estrela AB, Nascimento AK, Melo MM, Torres-Rêgo M, Lima EO, Rocha HA, Carvalho E, Silva-Junior AA, and Fernandes-Pedrosa MF. 2016. Characterization of TistH, a multifunctional peptide from the scorpion *Tityus stigmurus*: structure, cytotoxicity and antimicrobial activity. *Toxicon* 119:362–370.

Manuwar A, Dreyer B, Böhmert A, Ullah A, Mughal Z, Akrem A, Ali SA, Schlüter H, and Betzel C. 2020. Proteomic investigations of two Pakistani Naja snake venoms species unravel the venom complexity, posttranslational modifications, and presence of extracellular vesicles. *Toxins (Basel)* 12 (11):669.

Mebs D. 2002. Scorpions and snakes, such as cobras, mambas and vipers made the African continent famous for venomous animals. *Bulletin de la Societe de Pathologie Exotique* 95 (3):131.

Mendes LC, Viana GM, Nencioni AL, Pimenta DC, and Beraldo-Neto E. 2023. Scorpion peptides and ion channels: an insightful review of mechanisms and drug development. *Toxins* 15 (4):238.

Morey SS, Kiran KM, and Gadag JR. 2006. Purification and properties of hyaluronidase from *Palamneus gravimanus* (Indian black scorpion) venom. *Toxicon* 47 (2):188–195.

Mukherjee AK, Kalita B, and Mackessy SP. 2016. A proteomic analysis of Pakistan *Daboia russelii russelii* venom and assessment of potency of Indian polyvalent and monovalent antivenom. *Journal of Proteomics* 144:73–86.

Murthy KNC, Jayaprakasha GK, and Singh RP. 2002. Studies on antioxidant activity of pomegranate (*Punica granatum*) peel extract using in vivo models. *Journal of Agricultural Food Chemistry* 50 (17):4791–4795.

Narahashi T, Shapiro B, Deguchi T, Scuka M, and Wang CM. 1972. Effects of scorpion venom on squid axon membranes. *American Journal of Physiology-Legacy Content* 222 (4):850–857.

Orengo CA, Jones DT, and Thornton JM. 1994. Protein superfamilles and domain superfolds. *Nature* 372 (6507):631–634.

Ortiz E, Gurrola GB, Schwartz EF, and Possani LD. 2015. Scorpion venom components as potential candidates for drug development. *Toxicon* 93:125–135.

Oukkache N, El Jaoudi R, Chgoury F, Rocha MT, and Sabatier JM. 2015. Characterization of Am IT, an anti-insect β-toxin isolated from the venom of scorpion *Androctonus mauretanicus. Acta Physiologica Sonica* 25:295–304.

Patra A, Kalita B, Chanda A, and Mukherjee AK. 2017. Proteomics and antivenomics of *Echis carinatus carinatus* venom: correlation with pharmacological properties and pathophysiology of envenomation. *Scientific Reports* 7 (1):17119.

Pedarzani P, D'hoedt D, Doorty KB, Wadsworth JDF, Joseph JS, Jeyaseelan K, Kini RM, Gadre SV, Sapatnekar SM, Stocker M, and Strong PN . 2002. Tamapin, a venom peptide from the Indian red scorpion (*Mesobuthus tamulus*) that targets small conductance Ca^{2+}-activated K^+ channels and afterhyperpolarization currents in central neurons. *Journal of Biological Chemistry* 277 (48):46101–46109.

Peng F, Zeng XC, He XH, Pu J, Li WX, Zhu ZH, and Liu H. 2002. Molecular cloning and functional expression of a gene encoding an antiarrhythmia peptide derived from the scorpion toxin. *European Journal of Biochemistry* 269 (18):4468–4475.

Pless SA, and Ahern CA. 2015. Introduction: applying chemical biology to ion channels. In *Novel Chemical Tools to Study Ion Channel Biology*:1–4.

Polis GA. 1990. *The Biology of Scorpions*, edited by GA Polis. CABI Digital Library. Stanford: Stanford University Press.

Possani LD, Merino E, Corona M, Bolivar F, and Becerril B. 2000. Peptides and genes coding for scorpion toxins that affect ion-channels. *Biochimie* 82 (9–10):861–868.

Prendini L. 2000. Phylogeny and classification of the superfamily Scorpionoidea Latreille 1802 (Chelicerata, Scorpiones): an exemplar approach. *Cladistics* 16 (1):1–78.

Puillandre N, Watkins M, and Olivera BM. 2010. Evolution of Conus peptide genes: duplication and positive selection in the A-superfamily. *Journal of Molecular Evolution* 70:190–202.

Qiu S, Yi H, Liu H, Cao Z, Wu Y, and Li W. 2009. Molecular Information of charybdotoxin blockade in the large conductance calcium-activated potassium channel. *Journal of Chemical Information* 49:1831–1838.

Quintero-Hernández V, Jiménez-Vargas JM, Gurrola GB, Valdivia HH, and Possani LD. 2013. Scorpion venom components that affect ion-channels function. *Toxicon* 76:328–342.

Ramanaiah M, Parthasarathy PR, and Venkaiah B. 1990. Isolation and characterization of hyaluronidase from scorpion (*Heterometrus fulvipes*) venom. *Biochemistry International* 20 (2):301–310.

Rao VR, Perez-Neut M, Kaja S, and Gentile S. 2015. Voltage-gated ion channels in cancer cell proliferation. *Cancers* 7 (2):849–875.

Restano-Cassulini R, Garcia W, Paniagua-Solís JF, and Possani LD. 2017. Antivenom evaluation by electrophysiological analysis. *Toxins* 9 (3):74.

Rokyta DR, Wray KP, Lemmon AR, Lemmon EM, and Caudle SB. 2011. A high-throughput venom-gland transcriptome for the Eastern Diamondback Rattlesnake (*Crotalus adamanteus*) and evidence for pervasive positive selection across toxin classes. *Toxicon* 57 (5):657–671.

Romero-Gutiérrez MT, Santibáñez-López CE, Jiménez-Vargas JM, Batista CVF, Ortiz E, and Possani LD. 2018. Transcriptomic and proteomic analyses reveal the diversity of venom components from the vaejovid scorpion *Serradigitus gertschi*. *Toxins* 10 (9):359.

Romero-Gutierrez T, Peguero-Sanchez E, Cevallos MA, Batista CVF, Ortiz E, and Possani LD. 2017. A deeper examination of *Thorellius atrox* scorpion venom components with omic techonologies. *Toxins* 9 (12):399.

Rowan EG, Vatanpour H, Furman BL, Harvey AL, Tanira MOM, and Gopalakrishnakone P. 1992. The effects of Indian red scorpion *Buthus tamulus* venom *in vivo* and *in vitro*. *Toxicon* 30 (10):1157–1164.

Santibáñez López C, Cid-Uribe J, Batista C, Ortiz E, and Possani LD. 2016. Venom gland transcriptomic and proteomic analyses of the enigmatic scorpion *Superstitionia donensis* (Scorpiones: Superstitioniidae), with insights on the evolution of its venom components. *Toxins* 8 (12):367.

Santibáñez-López CE, Kriebel R, Ballesteros JA, Rush N, Witter Z, Williams J, Janies DA, and Sharma PP. 2018. Integration of phylogenomics and molecular modeling reveals lineage-specific diversification of toxins in scorpions. *PeerJ* 6:e5902.

Santos MS, Silva CG, Neto BS, Grangeiro Júnior CR, Lopes VH, Teixeira Júnior AG, Bezerra DA, Luna JV, Cordeiro JB, Júnior JG, and Lima MA. 2016. Clinical and epidemiological aspects of scorpionism in the world: a systematic review. *Wilderness Environmental Medicine* 27 (4):504–518.

Sato C, Ueno Y, Asai K, Takahashi K, Sato M, Engel A, and Fujiyoshi Y. 2001. The voltage-sensitive sodium channel is a bell-shaped molecule with several cavities. *Nature* 409 (6823):1047–1051.

Saviola AJ, Negrão F, and Yates III JR. 2020. Proteomics of select neglected tropical diseases. *Annual Review of Analytical Chemistry* 13:315–336.

Seyedian R, Pipelzadeh MH, Jalali A, Kim E, Lee H, Kang C, Cha M, Sohn ET, Jung ES, Rahmani AH, and Mirakabady AZ. 2010. Enzymatic analysis of *Hemiscorpius lepturus* scorpion venom using zymography and venom-specific antivenin. *Toxicon* 56 (4):521–525.

Shalit T, Elinger D, Savidor A, Gabashvili A, and Levin Y. 2015. MS1-based label-free proteomics using a quadrupole orbitrap mass spectrometer. *Journal of Proteome Research* 14 (4):1979–1986.

Sharma PP, Kaluziak ST, Pérez-Porro AR, González VL, Hormiga G, Wheeler WC, and Giribet G. 2014. Phylogenomic interrogation of Arachnida reveals systemic conflicts in phylogenetic signal. *Molecular Biology Evolution* 31 (11):2963–2984.

Shi J, He H, Zhao R, Duan Y, Chen J, Chen Y, Yang J, Zhang J, Shu Y, Zheng P, and Ji Y, 2008. Inhibition of martentoxin on neuronal BK channel subtype (alpha þ beta4): implications for a novel interaction model. *Biophysical Journal* 94:3706–3713.

Shichor I, Zlotkin E, Ilan N, Chikashvili D, Stuhmer W, Gordon D, and Lotan I. 2002. Domain 2 of *Drosophila* para voltage-gated sodium channel confers insect properties to a rat brain channel. *Journal of Neuroscience* 22 (11):4364–4371.

So WL, Leung TCN, Nong W, Bendena WG, Ming Ngai S, and Hui JHL. 2021. Transcriptomic and proteomic analyses of venom glands from scorpions *Liocheles australasiae*, *Mesobuthus martensii*, and *Scorpio maurus palmatus*. *Peptides* 146:170643.

Srairi-Abid N, Othman H, Aissaoui D, and BenAissa R. 2019. Anti-tumoral effect of scorpion peptides: emerging new cellular targets and signaling pathways. *Cell Calcium* 80:160–174.

Stevens M, Peigneur S, and Tytgat J. 2011. Neurotoxins and their binding areas on voltage-gated sodium channels. *Frontiers in Pharmacology* 2:71.

Stock L, Souza C, and Treptow W. 2013. Structural basis for activation of voltage-gated cation channels. *Biochemistry* 52 (9):1501–1513.

Strong PN, Clark GS, Armugam A, De-Allie FA, Joseph JS, Yemul V, Deshpande JM, Kamat R, Gadre SV, Gopalakrishnakone P, and Kini RM. 2001. Tamulustoxin: a novel potassium channel blocker from the venom of the Indian red scorpion *Mesobuthus tamulus*. *Archives of Biochemistry Biophysics* 385 (1):138–144.

Sunagar K, Johnson WE, O'Brien SJ, Vasconcelos V, and Antunes A. 2012. Evolution of CRISPs associated with toxicoferan-reptilian venom and mammalian reproduction. *Molecular Biology Evolution* 29 (7):1807–1822.

Tan NH, Fung SY, Tan KY, Yap MKK, Gnanathasan CA, and Tan CH. 2015. Functional venomics of the Sri Lankan Russell's viper (*Daboia russelii*) and its toxinological correlations. *Journal of Proteomics* 128:403–423.

Thakur R, and Mukherjee AK. 2015. A brief appraisal on Russell's viper venom (*Daboia russelii russelii*) proteinases. *Snake Venoms*:1–18.

Tobassum S, Tahir HM, Arshad M, Zahid MT, Ali S, and Ahsan MM. 2020. Nature and applications of scorpion venom: an overview. *Toxin Reviews* 39 (3):214–225.

Tobassum S, Tahir HM, Zahid MT, Gardner QA, and Ahsan MM. 2018. Effect of milking method, diet, and temperature on venom production in scorpions. *Journal of Insect Science* 18 (4):19.

Touyz RM, Alves-Lopes R, Rios FJ, Camargo LL, Anagnostopoulou A, Arner A, and Montezano AC. 2018. Vascular smooth muscle contraction in hypertension. *Cardiovascular Research* 114 (4):529–539.

Tytgat J, Chandy KG, Garcia ML, Gutman GA, Martin-Eauclaire MF, Van Der Walt JJ, and Possani LD. 2000. A unified nomenclature for short chain peptides isolated from scorpion venom: alpha-KTx molecular subfamilies. *Biophysical Journal* 78 (1):172.

Valdez-Cruz NA, Batista CVF, and Possani LD. 2004. Phaiodactylipin, a glycosylated heterodimeric phospholipase A$_2$ from the venom of the scorpion *Anuroctonus phaiodactylus*. *European Journal of Biochemistry* 271 (8):1453–1464.

Van Theemsche KM, Van de Sande DV, Snyders DJ, and Labro AJ. 2020. Hydrophobic drug/toxin binding sites in voltage-dependent K$^+$ and Na$^+$ channels. *Frontiers in Pharmacology* 11:735.

Vargas E, Yarov-Yarovoy V, Khalili-Araghi F, Catterall WA, Klein ML, Tarek M, Lindahl E, Schulten K, Perozo E, Bezanilla F, and Roux B. 2012. An emerging consensus on voltage-dependent gating from computational modeling and molecular dynamics simulations. *Journal of General Physiology* 140 (6):587–594.

Vatanpour H, Rowan EG, and Harvey AL. 1993. Effects of scorpion (*Buthus tamulus*) venom on neuromuscular transmission in vitro. *Toxicon* 31 (11):1373–1384.

Waddington J, Rudkin DM, and Dunlop JA. 2015. A new mid-Silurian aquatic scorpion—one step closer to land? *Biology Letters* 11 (1):20140815.

Weinberger H, Moran Y, Gordon D, Turkov M, Kahn R, and Gurevitz M. 2010. Positions under positive selection—key for selectivity and potency of scorpion α-toxins. *Molecular Biology Evolution* 27 (5):1025–1034.

Wong ESW, and Belov K. 2012. Venom evolution through gene duplications. *Gene* 496 (1):1–7.

Wudayagiri R, Inceoglu B, Herrmann R, Derbel M, Choudary PV, and Hammock BD. 2001. Isolation and characterization of a novel lepidopteran-selective toxin from the venom of South Indian red scorpion, *Mesobuthus tamulus*. *BMC Biochemistry* 2:1–8.

Xu X, Duan Z, Di Z, He Y, Li J, Li Z, Xie C, Zeng X, Cao Z, Wu Y, and Liang S. 2014. Proteomic analysis of the venom from the scorpion *Mesobuthus martensii*. *Journal of Proteomics* 106:162–180.

Yamashita T, and Rhoads DD. 2013. Species delimitation and morphological divergence in the scorpion *Centruroides vittatus* (Say, 1821): insights from phylogeography. *PLoS One* 8 (7):e68282.

Yi H, Qiu S, Cao Z, Wu Y, and Li W. 2008. Molecular basis of inhibitory peptide maurotoxin recognizing Kv1.2 channel explored by ZDOCK and molecular dynamic simulations. *Proteins: Structure, Function, Bioinformatics* 70:844–854.

Yin SJ, Jiang L, Yi H, Han S, Yang DW, Liu ML, Liu H, Cao ZJ, Wu YL, and Li WX. 2008. Different residues in channel turret determining the selectivity of ADWX-1 inhibitor peptide between Kv1.1 and Kv1.3 channels. *Journal of Proteome Research* 7:4890–4897.

Yu FH, and Catterall WA. 2003. Overview of the voltage-gated sodium channel family. *Genome Biology* 4 (3):1–7.

Zamudio FZ, Conde R, Arévalo C, Becerril B, Martin BM, Valdivia HH, and Possani LD. 1997. The mechanism of inhibition of ryanodine receptor channels by imperatoxin I, a heterodimeric protein from the scorpion *Pandinus imperator*. *Journal of Biological Chemistry* 272 (18):11886–11894.

Zeh DW. 1990. *The Biology of Scorpions*, edited by GA Polis. Vol. 249. Stanford, CA: Stanford University Press.

Zelanis A, and Tashima AK. 2014. Unraveling snake venom complexity with 'omics' approaches: challenges and perspectives. *Toxicon* 87:131–134.

Zeng X-C, Corzo G, and Hahin R. 2005. Scorpion venom peptides without disulfide bridges. *IUBMB Life* 57 (1):13–21.

Zeng X-C, Wang S-X, Zhu Y, Zhu S-Y, and Li W-X. 2004. Identification and functional characterization of novel scorpion venom peptides with no disulfide bridge from *Buthus martensii* Karsch. *Peptides* 25 (2):143–150.

Zhao Y, Chen Z, Cao Z, Li W, and Wu Y. 2019. Diverse structural features of potassium channels characterized by scorpion toxins as molecular probes. *Molecules* 24 (11):2045.

Zhang L, Shi W, Zeng X-C, Ge F, Yang M, Nie Y, Bao A, Wu S, and Guoji E, 2015. Unique diversity of the venom peptides from the scorpion *Androctonus bicolor* revealed by transcriptomic and proteomic analysis. *Journal of Proteomics* 128: 231–250.

Zhu W, Smith JW, and Huang C-M. 2010. Mass spectrometry-based label-free quantitative proteomics. *BioMed Research International* 2010:1–6.

Ziganshin RH, Kovalchuk SI, Arapidi GP, Starkov VG, Hoang AN, Nguyen TT, Nguyen KC, Shoibonov BB, Tsetlin VI, and Utkin YN. 2015. Quantitative proteomic analysis of Vietnamese krait venoms: neurotoxins are the major components in *Bungarus multicinctus* and phospholipases A$_2$ in *Bungarus fasciatus*. *Toxicon* 107:197–209.

4 Scorpionism
Epidemiology, Pathophysiology, and Clinical Manifestations

4.1 FACTORS AFFECTING THE SEVERITY OF SCORPION STING IN A REGION

Scorpions originated about 450 million years ago and (Sharma et al. 2014; Lourenço 2018) and one million scorpion stings occur in all continents (except Antarctica), although the epidemiology of scorpion stings is less known. It thus remains a serious health problem worldwide (Bahloul et al. 2010; Chipax and Goyfon 2008). Prevention methods aimed at lowering the prevalence of scorpion sting cases are significantly impacted by the high number of stings that occur all year long (Albuquerque et al. 2013). Therefore, it is essential to understand the epidemiology of this injury (Barbosa et al. 2012). The severity of a scorpion sting depends on the scorpion species and the victim.

Clinical symptoms of a scorpion sting in a particular locality depend on several significant factors. The dimensions of the scorpion, the amount and type of toxins in its venom, the location of the telson, the venom duct, the number of stings, and the victim's age and health are some of these variables (Santos et al. 2016; Tiwari and Deshpande 1993). It is important to note that children and adults with weakened immunity are particularly vulnerable to this type of disaster (Dabo et al. 2011; Goyffon and Billiald 2007; Matthiensen 1988; Soares et al. 2002).

The seasonal patterns in the temporal incidence of scorpion stings are highly associated with climatological factors. The frequency or intensity of scorpion stings varies significantly by location (Bahloul et al. 2010; Chippaux and Goyffon 2008). Significant regions for the frequency and intensity of scorpion stings include the Near and Middle East, Mexico, Brazil, South and East Africa, North-Saharan Africa, the Amazonian Basin (which consists of Northern Brazil, the Guyanas, Venezuela), and South India (Chippaux and Goyffon 2008). Scorpion stings can cause serious health problems such as heart failure, pulmonary failure, hyperglycemia, hypertension, and other local and systemic symptoms. In tropical and subtropical areas, they are considered life-threatening medical emergencies (Bawaskar and Bawaskar 1992; Das et al. 2021; More et al. 2004; Kularatne et al. 2015).

DOI: 10.1201/9781003540816-4

4.2 IMPORTANCE OF UNDERSTANDING THE EPIDEMIOLOGY OF SCORPIONISM: POTENCY AND SEVERITY

As already mentioned, scorpion stings and their effects are a significant cause of emergency, particularly among children (Almitai et al. 1985; Dehesa-Dávila and Possani 1994; El-Amin et al. 1994; Goyffon et al. 1982; Lourenço et al. 1996; Otero et al. 2004). Scorpionism is a severe public health issue in many parts of the world, as its envenomation is often painful and lethal to humans and causes severe health complications (Bawaskar and Bawaskar 2012; Chippaux 2015; Khatony et al. 2015; Maghsoodi et al. 2015; Ebrahimi et al. 2017). As already stated, the incidence and severity of scorpionism vary considerably according to geographical location. Thus, it is essential to identify the local population's specific concerns and significant risk factors.

Scorpionism is significant on varying levels in seven different continents of the world. In endemic urban settings, epidemiological population-based study on scorpions or investigations into the features of the etiological agents responsible for the events are required (Bahloul et al. 2010; Chippaux and Goyffon 2008)

Because of the high number of stings that occur all year long, preventative methods have been developed to lower the prevalence of scorpion stings (Albuquerque et al. 2013). Thus, it is crucial to understand the epidemiology of this injury (Barbosa et al. 2012). There are 2,231 species of scorpions worldwide, belonging to 208 genera and 20 families (Zhang et al. 2015; Romero Gutierrez et al. 2017). Of these, 1,500 species are venomous, and about 50 are exceptionally dangerous; therefore, they are hazardous for humans. Most of these species are found in the genera *Buthus*, *Parabuthus*, *Mesobuthus*, *Tityus*, *Leiurus*, and *Androctonus*, belonging to the Buthidae family (Uluğ et al. 2012; Özkan and Karaer 2003; Srinivasan et al. 2002; Gomes and Gomes 2015; Gomes et al. 2016; Ebrahimi et al. 2017).

Scorpions of the Buthidae family are more lethal and medically significant than many other scorpion families (Quintero-Hernandez et al. 2013; Ortiz et al. 2015). Their venom is vital to their survival by protecting them from competitors, predators, and intruders. Every year, there are more than 1.2 million scorpion stings, 3,000 deaths from scorpion envenomation, and over 2.3 billion people reside in places where there is a risk of being bitten by a scorpion (Chippaux 2012; Bahloul et al. 2013; Khatony et al. 2015; Ebrahimi et al. 2017). However, low socioeconomic status and limited health facilities lead to high mortality rates from scorpions in many poor and developing countries (Natu et al. 2006). The probability of venomation by a scorpion is contingent upon multiple factors about both the victim and the scorpion (Dabo et al. 2011; Goyffon and Billiald 2007; Matthiensen 1988; Soares et al. 2002).

4.3 EPIDEMIOLOGY AND CLINICAL SYMPTOMS OF SCORPION STINGS AROUND THE WORLD

In several tropical countries, scorpion sting is a significant public health issue due to its frequent occurrence and potential severity. Despite the wide prevalence and risk, epidemiological data on scorpion disease still needs to be expanded due to the lack of unreported instances and studies on this issue (Chippaux and Goyffon 2008;

Cesaretli and Ozkan 2010). The risk of encountering scorpions has grown in many of these regions due to environmental change brought about by population development and poorly maintained urban health facilities. In contrast, the sphere's northern half—including the USA, Canada, Europe, Russia, and Australia—is not home to the deadliest scorpions.

As scorpionism frequently affects infants and the elderly, whose immune systems are more vulnerable and have been linked to higher death rates and lethality, the public and governments should take persistent action against it. The sympathetic and parasympathetic nervous systems are implicated in the genesis of clinical symptoms. The excitatory neurotoxins found in scorpion venom cause a massive release of neurotransmitters responsible for autonomic storms (adrenergic or cholinergic autonomic storms). This opening of calcium channels at presynaptic nerve terminals triggers an autonomic storm.

4.3.1 EPIDEMIOLOGY OF SCORPIONISM IN AMERICA

4.3.1.1 Epidemiological Study in South America

The majority of South American scorpion species that kill humans belong to the genus *Tityus*. Studies investigating the features of the etiological factors producing the incidents in metropolitan regions where this problem is prominent are scarce, as is epidemiological population-based research on scorpions.

In Brazil, a country in South America, scorpionism is becoming more significant due to increased envenomations and fatalities, mostly in metropolitan areas. Brazil has around 160 species of scorpions, but the genus *Tityus* is responsible for substantial stings from a medical perspective—four *Tityus* species, viz. *Tityus serrulatus*, *T. bahiensis*, *T. stigmurus*, and *T. obscurus* are of medicinal interest (Ministério da Saúde 2009). The *T. serrulatus* is medically most significant because it is responsible for many regional stings. The patient's venom composition and physical state determine the severity of the systemic effects of *Tityus* species. Adults with comorbidities and children under 12 years of age are typically the most severely impacted, and most fatalities have occurred in these age ranges (Ministério da Saúde 2018; Cupo 2015; Reckziegel and Pinto 2014).

The hazardous scorpion encounters in Brazil alone increased from 64,000 to 124,000 annually in 2012 (Reckziegel and Pinto 2014). Most human-deadly scorpion species are members of the Buthidae family (Laustsen et al. 2016). However, several species in the families of Hemiscorpidae and Scorpionidae are also considered invasive (Lourenço 2018; Hauke and Herzig 2017). The distribution of those medicinally significant species across geography is linked to the domestic expansion of scorpions. Thus, there has been a notable increase in the intensity of severe species in Northern Africa, Brazil, Mexico, Saudi Arabia, Iran, and Venezuela (Chippaux 2012; Mullen and Durden 2019; Ward et al. 2018).

The number of documented deaths in Brazil from scorpion stings has increased over the last 11 years, from 61 in 2007 to 90 in 2017, according to the country's public health system. During this time, there has also been an increase in scorpion sting cases, from 37,370 to 124,982. About 83% of deaths in the past 5 years (2013–2017) happened within 48 hours of being stung (Ministério da Saúde 2018). The fatality rate is often less than 0.09% in other age groups (between 20 and 50 years), although 0.32% and 0.13%, respectively, for victims under 10 years and over 75 years. However, the fatality rate for children between 1 and 5 years of age is 0.40 % (Ministério da Saúde 2018).

The rapid urbanization witnessed in Brazil over the past few decades has created the perfect environment for spreading opportunistic and invasive scorpions like *T. serrulatus* and *T. stigmurus*. These scorpions are invasive because there has not been enough development of basic infrastructure, including water, light, wastewater treatment, and regular waste collection. Being parthenogenetic species, both can reproduce without the male's fertilization. Both have remarkable flexibility in adjusting to human habitations and are suited to dramatically altered settings, as in various cities in Brazil where human populations have increased.

The above condition is prevalent in some parts of Brazil, Mexico, and North Africa (von Eickstedt et al. 1996; Lourenço et al. 1996; Bertani et al. 2018). Most severe illnesses and fatalities are caused by the sting of *T. serrulatus* (Cupo 2015; Reckziegel and Pinto 2014). Although its origin is uncertain, previous documents point to the Brazilian state of Minas Gerais. Species identification, whether by capture or photography, is crucial, especially if it is one of the four *Tityus* spp. with significant medical value. *T. obscurus* needs to be recognized in the Amazon region as it can potentially induce acute cerebellar impairment (Cupo 2015; Reckziegel and Pinto 2014; Torrez et al. 2015). The Brazilian Association of Information Centers and Toxicological Assistance (ABRACIT), which is made up of public institutions, offers free help and support regarding envenoming management in cities where the antiscorpion antivenom (ASA) is available in every state hospital, as well as the closest Reference Health Unit with ASA, via phone or the Internet (Ministério da Saúde 2018; Azevedo 2006).

According to the degree of the envenoming, the symptoms followed by scorpion stings in Brazil can be divided into three phases. The first envenomation is often benign and progresses to stage I, distinguished by severe pain (stage Ia), stirring, fever, sweating, nausea, and changes in blood pressure (stage Ib). In extreme situations (5–10% of cases), stage II progresses from stage I and is characterized by sweating, vomiting, cramps, diarrhea, hypotension, bradycardia, pulmonary obstruction, and dyspnea. Stage III is the final and most dangerous stage, characterized by respiratory distress such as pulmonary edema, bronchospasm, and cyanosis, as well as the possibility of hyperthermia, cardiac arrhythmia, and myocardial ischemia (Bahloul et al. 2004; Chippaux and Goyffon 2008). Although the severity of scorpion envenomation varies depending on the species, age, and size of the scorpion, it is substantially worse in children (Amitai 1998). Treatment options for scorpion stings include symptomatic treatment, assistance with necessary physiological processes, and, in extreme circumstances, ASA therapy (Abroug et al. 2003).

A study revealed that 91.7% of the victims had symptoms and local indications in Brazil, while 98.6% of the envenomation patients had systemic manifestations (Pardal et al. 2003). However, according to another study from Brazil, scorpion stings can have local and systemic effects (Barbosa et al. 2012). However, the former clinical symptom was more prevalent (86.2%). Only 8.99% of the victims in this study from Brazil showed systemic symptoms with mild to severe clinical signs, and 1.85% of the victims who obtained medical attention had no symptoms. In the research by Bahloul et al. (2010), 555 patients (81%) experienced systemic inflammatory response syndrome, and 552 patients (80.6%) experienced organ failure in at least one organ. There were 542 patients (79.1%) who had cardiac failure, 444 (64.8%) experienced respiratory failure, 25 (3.6%) suffered neurological signs and symptoms, 13 (1.9%) encountered renal failure, and 7 (1%) patients experienced hematologic or hepatic failure. More than two organs were found failing in most patients (64.7%).

In another research, liver failure was noted in 17 patients (1.8%) and was linked to an inferior prognosis. The death rate in the liver failure group was 41.1%, compared to 7.4% in the control group (P < 0.0001). Bahloul et al. (2010) reported from a university hospital in Tunisia that 26.4% of patients had cyanosis, and 80.2% had priapism. However, Adiguzel et al. (2007) noted cyanosis in just 1.8% of patients (report from Sanliurfa, Turkey). In addition, a study by Bosnak and colleagues found that in addition to priapism in children (17.3%), people concurrently exhibited other symptoms, such as autonomic storm (38.4%) and chilly extremities (38.4%).

When Pardal and his colleagues (1999) investigated scorpionism in a state in northern Brazil, they discovered that neurological signs were more prevalent than general symptoms. However, myoclonus (93%), dysmetria (86.1%), dysarthria (80.6%), and gait ataxia (70.8%) were the most prevalent symptoms, but electric shock feeling in the body was the more common symptom (88.9%). Another piece of data provided by Bahloul et al. (2010) showed that 580 patients (84.7%) experienced neurological symptoms. Consciousness abnormalities (Glasgow Coma Scale) were found in 177 patients (25.8%), 123 of whom (18%) were in coma (Glasgow Coma Scale). Nine individuals underwent a brain computed tomography scan; seven had cerebral edema or ischemia. The authors also provided further details on neurological symptoms such as squint (16%), bilateral myosis (5.7%), bilateral mydriasis (2%), and anisocoria (0.6%). De Sousa and their colleagues (2005) in Venezuela reported 21 deaths related to scorpions, with the following listed as direct causes: disseminated intravascular coagulation with bilateral pulmonary hemorrhage (4.8%), acute pulmonary edema (42.9%), acute pulmonary edema with congestive heart failure, and cardiac arrhythmias (19%); however, some symptoms are not specified (33.3%) (Santos et al. 2016).

According to Albuquerque and his colleagues (Albuquerque et al. 2013), most scorpion envenomation cases in Brazil were successfully treated. But during the research period, three toddlers under the age of 5 died from scorpion stings. Similarly, two children, aged 4 and 6, died. Soares and his colleagues found that 1994 had the highest mortality toll, with four male deaths (1.04%). This study was conducted in Brazil. Mejas and his colleagues claimed that scorpionism was the reason for 11 deaths in Venezuela between 1995 and 2000, where 8.9% of deaths were noted (Bahloul et al. 2010).

The Brazilian scorpion *T. serrulatus* caused increased lung, liver, kidney, and heart tenderness in rats post injection, defined by an increased density of mononuclear cells (Peres et al. 2009). The scientists concluded that the venom causes acute lung damage, including altered lung function and increased pulmonary inflammation. The primary pathogenic mechanism is pulmonary capillary endothelial failure, which causes exudative fluid rich in phagocytic immune cells and interstitial and alveolar edema (Amaral et al. 1994; Amaral and Rezende 1997). After being stung, pulmonary edema may appear suddenly. The signs and symptoms of pulmonary edema vary, although they can occur quickly. Depending on the severity and length of the renal failure, an accumulation is followed by problems such as metabolic acidosis and hyperkalemia, changes in blood fluid balance, and effects on multiple other organ systems. Renal edema significantly increased due to the scorpion venom's decreased glomerular filtration rate and urine flow.

According to de Sousa Alves et al. (2005) and Severino et al. (2009), the *T. serrulatus* venom also has an impact on hemodynamics by increasing renal flow in mice directly by a direct vasoconstrictor effect (de Sousa Alves et al. 2005; Severino et al. 2009)—acute renal failure results from a sudden loss of kidney function due to kidney injury. According to several investigations, scorpion venom significantly increases renal edema, connected to a lower glomerular filtration rate and urine flow. According to reports, scorpion stings can lead to abrupt renal failure (Pessini et al. 2008; Mahadevan 2000; Radmanesh 1990). In a healthy condition, the kidney maintains glomerular filtration and renal blood flow by autoregulation that is dependent on the tone of the afferent and efferent arterioles. The relative hypovolemia of sepsis causes renal hypoperfusion and cytokine-induced systemic vasodilation (Sofer and Gueron 1990).

4.3.1.2   Epidemiological Study in North America

In Guanajuato, Mexico, a country in North America that is home to the venomous scorpion species *Centruroides infamatus infamatus*, there has been an increase in scorpion activity during the warmer months (Dehesa-Dávila and Possani 1994). Similar observations were made in Argentina, where the warmer months of October through April saw an increase in the prevalence of *Tityus trivittatus* scorpion stings (de Roodt et al. 2003). The scorpion sting cases decrease at the beginning of the rainy season (Dehesa-Dávila 1989; Chowell et al. 2005). Most scorpions spend the day hiding in caves, leaf litter, or rocks. Nonetheless, research by Chowell and associates (Chowell et al. 2005) shows a substantial and favorable correlation between Mexico's lowest temperature and scorpion activity. Scorpion stings were reported most common in 2000 and 2001 during the lowest temperatures of 19.4°C and 18.8°C, respectively. This correlation is consistent with research conducted in Argentina and Brazil (Dehesa-Dávila and Possani 1994; de Roodt et al. 2003). Additionally, scientists noted a "threshold" relationship between the frequency of scorpion stings and pluvial precipitation (Chowell et al. 2005).

Scorpion stings were rare when rainfall was lower than 30 mm/month. Nonetheless, rainfall exceeding 30 mm/month did not affect the frequency of scorpion stings. The probable reason could be that rain disturbs the scorpions, forcing them to find new shelter. Several active scorpion venom components that act on Ca^{2+} and Cl^- channels have also been described (Guéguinou et al. 2014; Possani et al. 2000; Rao et al. 2015; Simms and Zamponi 2014). Children in Arizona frequently experience severe and perhaps fatal *Centruroides* scorpion envenomation. The presentation may mimic a seizure or anaphylaxis to a less-experienced doctor. Young patients with symptoms severe enough to necessitate assessment at a medical facility typically have respiratory impairment (Bond 1992). In the present era of pediatric critical care, deaths from scorpion stings have not been documented, but morbidity may be severe (Baseer and Naser 2019).

The most severe respiratory complication in scorpion sting victims is pulmonary edema. Most clinical and experimental studies suggest a hemodynamic origin for this condition, despite some clinical findings linking pulmonary edema to enhanced capillary permeability. Acute left ventricular failure brought on by a massive release of catecholamines or myocardial injury brought on by venom has been related to acute pulmonary edema. The origin of ventricular dysfunction and myocardial damage caused by severe scorpion envenomation has been the subject of much discussion in the past (Oswald et al. 1998; Sedziwy, Thomas, Shilliigford 1968). The initially recognized theory put up the higher blood catecholamine levels due to the venom's direct stimulation of the sympathetic nervous system and adrenal glands. The second acknowledged theory postulates about the venom's immediate impact on the heart, which results in "scorpion myocarditis" (Bahloul et al. 2013). It is likely that the venom directly affects the cardiac cell membranes, changing their permeability and electrical characteristics, leading to functional impairments through aberrant electrolyte fluxes and shifts. The final recognized theory is that myocardial ischemia may cause cardiac dysfunction. The last recognized approach is that myocardial ischemia may cause cardiac dysfunction (Mishra and Prasad 2015; Chippaux 2012).

4.3.2   Epidemiology of Scorpionism in Europe

France and its territories have documented 67 species of scorpions with extensive distribution. The number of stings may be more consistent, up to 90 per 100,000 people annually. Superficial visual characteristics are widely used to identify scorpion species, including species of medicinal importance (e.g., *Buthus, Centruroides*, and *Tityus*) (Vaucel et al. 2022).

4.3.3   Epidemiology of Scorpionism in Asia

Some East Asian countries (such as China and Mongolia), South Asia (such as India, Pakistan, and Sri Lanka), and West Asia (Iran, Turkey, and Saudi Arabia) show high incidence of scorpion stings (Shahi et al. 2015; Ebrahimi et al. 2017). In Iran, scorpion stings predominantly occur in three provinces, viz. Khuzestan, Bueyerahmad, and Kohgiluyeh (Maghsoodi et al. 2015; Shahi et al. 2015).

4.3.3.1   Epidemiological Study in South Asia

Scorpion epidemiology and regional distribution have been the subject of much investigation. However, this is not the case in Pakistan, where it is still not clear how common scorpion stings are due to a lack of published documented data. In Pakistan, two epidemiological studies on scorpion stings have been published: one from the Lasbella District in Balochistan (Khan and Ullaha 2017) and the other from the District Sargodha in Punjab (Ahsan et al. 2015).

An epidemiological survey between 1984 and 1995 revealed that children from Sri Lanka who had been stung by scorpions were discovered in a place called Mahad, which is 200 km south of Mumbai, Western India. Of these children, 13 (57%) were male and 10 (43%) were female. Upon admission to the hospital, every patient displayed symptoms of envenoming, whether localized or systemic (Kularatne et al. 2015; Ratnayake et al. 2016).

It was found that in the northern part of Sri Lanka, particularly in the Jaffna district, scorpion stings were seriously harming people. Not until 2009 was the northern region of Sri Lanka accessible to medical researchers because of the 30-year-long war that has devastated the area. Therefore, until doctors at the Teaching Hospital Jaffna (THJ) brought the problem of scorpion stings to the attention of researchers on invasive species of scorpions, it remained unaddressed. Locals and medical experts called the aggressive scorpion the "White Scorpion" because they thought its color was paler than that of the "Black Scorpions" that are frequently encountered in the nation (primarily heterogametic species, which are known to have a considerable number of forest scorpions). The core peninsula and several islands (islets) on the western side comprise the 1114 km^2 Jaffna Peninsula. No documented material or firsthand accounts of "White Scorpion" stings are available throughout the pre-war and wartime eras (Kularatne et al. 2015).

Three locations in Jaffna, on the mainland, Achchuvali, Palali, and Karainagar, have been noted as scorpion sting hotspots. Thirteen (57%) and 10 (43%) of the confirmed cases had a mean age of 30 years (SD ± 20 years), with males and women making up the respective groups. Five (22%) of them were younger than 12. Within an hour and 3 hours, 13 (57%) and 18 (78%) reached the hospital, respectively. All patients reported localized or systemic symptoms of scorpion envenomation at the time of admission (Kularatne et al. 2015).

In three distinct locations in Jaffna, Sri Lanka, there have been 23 recorded stings by Indian red scorpion (*Mesobuthus tamulus*). Of the clinical reports, there were 13 (57%) males and 10 (43%) females, 5 (22%) cases involved children under the age of 12, and all patients reported displaying either local or systemic manifestations (Kularatne et al. 2015; Ratnayake et al. 2016).

In India, Kerala, Tamil Nadu, Andhra Pradesh, Karnataka, Saurashtra, and the western states of Maharashtra are regularly affected by morbidity and mortality brought on by scorpion stings. A clinical case study of 141 kids who were stung by an Indian red scorpion and brought to the Government Raja Mirasdhar Hospital in Thanjavur, southern India, revealed that kids between the ages of 1–3 and 7–12 had the worst reactions to envenomation. Five out of the eight people who had priapism were older than 6 years. Pulmonary edema was observed in one patient older than 6 years following the fatal and life-threatening sting effect (Yuvaraja et al. 2019).

Fifty patients in southern India were stung by an Indian red scorpion, according to epidemiological data from a tertiary care and teaching hospital. Patients showed symptoms of dyspnea (13, 26%), chest discomfort (9, 18%), vomiting (6, 12%), sweating (5, 10%), nausea (3, 6%), priapism (7, 14%), and piloerection (6, 12%) (Madhavan 2015).

While the coastline regions and the nearby islands have limited vegetation on sandy soil, the district's central area is rich in highly red soils. In 1 year, 90 people with a history of scorpion stings were hospitalized. *M. tamulus* was the offending scorpion in 84 cases, and black scorpions primarily stung others. Twenty-three patients or relatives could identify the offending scorpions as *M. tamulus*. In the remaining 61 cases, the victims or witnesses had sightings of the offending scorpions but could not capture them. There was no significant variation in clinical features between these two groups (Kularatne et al. 2015).

Other areas of India include Western Maharashtra, Tamil Nadu, Kerala, Saurashtra, Andhra Pradesh, and Karnataka. According to the study, children between 1–3 and 7–12 years had the worst outcomes. Of them, eight patients and five less than 6 years old demonstrated priapism. Pulmonary edema in a 6-year-old child was a severe and life-threatening clinical event (Yuvaraja et al. 2019). Having been stung by an Indian red scorpion, 50 patients in a southern Indian teaching and tertiary care hospital reported chest pain (9, 18%), dyspnea (13, 26%), vomiting (6, 12%), sweating (n=5, 10%), nausea (n=3, 6%), priapism (n=7, 14%), and piloerection (n=6, 12%) (Madhavan 2015). In Mahad (200 km south of Mumbai, Western India), an epidemiological study was carried out between 1984 and 1995 with 293 patients, of which 6 deaths were reported before hospital admission (Bawaskar and Bawaskar 1998).

Based on the clinical symptoms, patients were further divided into three broad groups:

1. Hypertension was observed in 111 (38%) patients within 1–10 hours (mean 3.5 hours).
2. Tachycardia was observed in 87 (30%) patients within 1–24 hours (mean 6.7 hours).
3. Pulmonary edema was observed in 72 (24.5%) patients within 6–24 hours (mean 8 hours) following a scorpion sting.

Following an Indian red scorpion sting on his right big toe, a 14-year-old young person (male) from the Babaganj area of Northern India developed gastrointestinal and cardiovascular issues (Agrawal et al. 2015). At the Rims Teaching Hospital in Raichur, Karnataka, India, 33 cases of scorpion stings were received between 2009 and 2014. Of them, 22 were caused by the Indian black scorpion and 11 by the red scorpion. The patients experienced hypotension or hypertension, bradycardia, cutaneous symptoms, and drowsiness (Rajashekhar and Mudgal 2017; Rajarajeswari et al. 1979).

Geographical variations in sting intensity have been reported in India (Reddy et al. 2017; Suranse et al. 2019). These variations are likely due to population genetic structure variation, which is thought to be the primary cause of phenotypic variations in venom composition (Newton et al. 2007). When populations of Indian red scorpions are gathered from eight locations in Maharashtra (Bhate plateau, Sangameshwar, Jejuri, Shindavane, Pashan, Alandi, Kalyan, and Jalna), a regression study shows that the genetic distance of subspecies increases by 0.006% (95% CI 0.003 to 0.010) per kilometer of distance from one another (Suranse et al. 2017).

Additionally, it has been proposed that variations in venom phenotype are correlated with variations in genetic structure and climatic precipitation, as well as places with extremely high, moderate, and low rainfall (Suranse et al. 2017). For example, there was a significant differential in the expression of venom peptide toxins between Indian red scorpions captured from the semi-arid Deccan plateau and the Konkan region of Maharashtra (Newton et al. 2007). Additionally, anecdotal evidence suggests that Indian red scorpion stings in the Konkan area on the western side of the Western Ghats are more painful than those in communities on the eastern side. These changes in the two populations' venom peptide makeup are probably to blame for these variances (Newton et al. 2007).

Additional factors might influence the sting's pathophysiology, but they have yet to be covered in the published literature (Bawaskar and Bawaskar 1999; Kankonkar et al. 1998; Murthy and Zare 1998; Newton et al. 2007). Analytical techniques such as sodium dodecyl–polyacrylamide gel electrophoresis (SDS–PAGE) have also been used to establish intra-specific venom variation between Indian red scorpions from Chennai and Western India (Ratnagiri, Chiplun, and Ahmednagar) (Badhe et al. 2006). Indian red scorpion venom obtained from the aforementioned regions was injected into mice at equivalent doses, yet the mice's blood sodium levels varied significantly (Badhe et al. 2007). Although comprehensive investigations of the venoms of Indian red scorpions across the Indian subcontinent are presently unachievable, these findings may provide insight into how venom composition is influenced by topography.

4.3.3.2   Epidemiological Study in West Asia

Scientists worldwide have long been interested in the scorpion fauna in Iran from the perspectives of systematics, biology, and ecology. The evaluation of species distribution data is based on research published in the scientific literature up to 2012. The Buthidae family, which includes 88.5% of all species and 82% of all genera among the 51 species of scorpions found in Iran's various areas, is home to most of the country's scorpion species. Scorpion stings have been reported nationwide (Dehghani et al. 2012). Shahbazzadeh and his team (Shahbazzadeh et al. 2009) reported that three patients in Iran died after being stung by scorpions. One such instance was a 13-year-old girl who was bitten by *Hemiscorpius lepturus* and suffered from severe heart failure and localized necrosis in her hand.

The *Androctonus* genera contain the most recognized medically relevant species within the Buthidae family. Other families of scorpions that are distributed in Iran are Hemiscorpiidae and Scorpionidae. Three species (5.75%) and two genera (9%) exist in the Hemiscorpiidae family. The medically dominant genus in this family, with the majority of notorious species, is *Hemiscorpius*—lastly, the Scorpionidae family has three species and two genera. The most populated areas of Iran are in the south and southwest, where about 95% of scorpion species are found (Navidpour et al. 2008.

There have been reports on the medical significance, epidemiology, and geographic distribution of scorpions in Iran (Dehghani 1998; Navidpour et al. 2008; Dehghani et al. 2012). Among the provinces of Iran, Khuzestan is the most prominent in terms of scorpions and scorpion stings (Vazirianzadeh et al. 2012). With 19 types of scorpions, Khuzestan is one of southwest Iran's most hazardous regions for scorpion stings. Due to the importance of scorpion stings and the lack of epidemiological information on this concern in public health, a study was conducted to gather fresh data on scorpion stings in Iran. Iran has different types of weather, including summer, cold, and snowy winters. One study reported that between 2002 and 2011, there were 55–66 scorpion stings per 100,000 people. These differences were likely caused mainly by diverse climatological factors and preventative strategies. Around 3,250 deaths are predicted to occur worldwide each year, affecting 1.2 million individuals. According to Chippaux and Goyffon (2008), the global mean rate of sting occurrences per 100,000 people yearly is approximately 17.14 (Chippaux and Goyffon 2008). It demonstrates that Iran experiences higher scorpion stings than the global average. According to research conducted in Kashan, central Iran (Dehgani et al. 2010), and Ahvaz, southwest Iran, scorpion stings were most common among people aged under 15–34 years (Emam et al. 2008). The highest incidence of scorpion sting cases (44.16%) in 2011 occurred in summer. This fact is consistent with findings from Iran (Rafeeazadeh 2009), Saudi Arabia (Shahbazzadeh et al. 2009), and Turkey (Jarrar and Al-Rowaily 2008; Dehgani et al. 2010; Dehgani and Fathi 2012; Emam et al. 2008; Ozkan and Kat 2005). According to their data, 49.7–93.4% of scorpion sting incidents happened in the summer. The severity of envenoming depends on the variability of scorpion venoms. Therefore, identifying the species that stings is crucial; this diagnosis will improve treatment.

Saudi Arabia also experienced scorpion stings; however, the mortality rate is low. At least 28 scorpion species have been discovered in Saudi Arabia. Each year, around 14,500 scorpion stings are recorded in various locations in Saudi Arabia (Alhamoud et al. 2021). The Buthidae family of scorpions, which includes *Androctonus crassicauda*, *Leiurus quinquestriatus*, *Mesobuthus gibbosus*, and *Mesobuthus eupeus* species, are the most dangerous in Turkey.

The epidemiological and clinical findings of scorpion stings in Turkey were studied through data from the National Poison Information Centre (NPIC) between 1995 and 2004; interestingly, most incidents occurred in July (Cesaretli and Ozkan 2010). A total of 930 examples were documented. The gender division was 50.22% female and 45.48% male. The age range of 20–29 years accounted for the majority of scorpion-stung cases. Central Anatolia and the Marmara regions in Turkey saw the most stings. Although there was no reported death, patients at the hospital showed signs of both localized (pain, edema, hyperemia, and numbness in the bitten part) and systemic consequences (hyperthermia, nausea and vomiting, tachycardia, shivering, and exhaustion). According to data, 33% of the envenomated individuals were treated in hospitals using ASA (Cesaretli and Ozkan 2010). According to Ulu et al. (2012), the majority of electrocardiographic changes in their series (in Turkey) were not sinus tachycardia, ST segment abnormalities, or extended QT intervals, which were observed in six (6.1%) and one (%) instances, respectively. In another Turkish investigation, Cesaretli and Ozkan noted that localized symptoms were observed in 77.9% of individuals (Cesaretli and Ozkan 2010).

Depending on the type of scorpion and the patient's age, gastrointestinal symptoms might occur anywhere between 11% and 100% of the time (de Roodt et al. 2003;

Isbister et al. 2003; Bouaziz et al. 1996; Bucaretchi et al. 1995). However, Bahloul and his team discovered that 507 (74%) patients developed gastrointestinal problems following scorpion stings in Tunisia. Death was much more significant in this group (P ¼.023), with 36 patients (5.3%) reporting having diarrhea, 24 patients (2.5%) reporting nausea, 41 patients (4.3%) having diarrhea, and 687 patients (72.2%) having vomiting (Bahloul et al. 2005). The sinus tachycardia was the most common abnormality post-scorpion stings (Bahloul et al. 2010). One hundred and thirty (19%) individuals had T-wave alterations on their electrocardiogram (ECG). In addition, 94 (13.7%) individuals had ST segment depression or elevation on their ECG, another abnormality. Additionally, in three patients who underwent echocardiography and scintigraphy, a low left ventricular ejection fraction (less than 45%) corroborated the cardiogenic nature of the pulmonary oedema.

In Tunisia, Nouira et al. (Nouira et al. 2007) proposed a simple score including seven crucial clinical risk factors of hospitalization, while Krifi et al. (Krifi et al. 1998) used three grades (Grade I: local signs, Grade II: general clinical symptoms without vital failure, Grade III: critical failure). Early triage of patients experiencing scorpion envenomation may benefit from it. Goyffon used three stages, but Soulaymani-Bencheikh et al. (Soulaymani-Bencheikh et al. 2002, 2008) utilized three classes (Class I: local signs limited to the sting region, Class II: widespread signs, and Class III: vital failure) to express the severity of scorpion stings.

4.3.3.3 Epidemiological Study in East Asia

Beginning with the first description of a Chinese scorpion (*Buthus martensii Karsch*, also known as *Mesobuthus martensii Karsch*) in 1879, the visiting scientists launched a taxonomic study of Chinese scorpions. Fifty-three species of scorpions belonging to 12 genera and 5 families (Buthidae, Chaerilidae, Euscorpiidae, Hemiscorpiidae, and Scorpionidae) have been reported from China. The Buthidae family, described by C. L. Koch in 1837, is the most widely distributed in China. It consists of 6 genera, 18 species, and subspecies, including 2 species each in *Hottentotta*, 3 species in *Isometrus*, 2 species each in *Lychas*, 9 species in *Mesobuthus*, 1 species in *Orthochirus*, and 1 species in *Razianus*. Except for Guizhou, Heilongjiang, Hunan, Jiangsu, Jilin, and Sichuan, they are dispersed over the majority of the provinces of China. No epidemiological studies of scorpion stings have been reported from China.

4.3.4 EPIDEMIOLOGY OF SCORPIONISM IN AFRICA

Despite the diversity of scorpion species found in South Africa, statistical data on the frequency and severity of scorpion envenomation is limited. A study was conducted with 52,163 consultations in the Tygerberg Poisons Information Centre (TPIC) of South Africa. They reported 740 (1.4%) cases involved in scorpion stings. Of them, 146 (19.7%) were considered significant envenomations. In these cases, adults (>20 years) made up 71.4% of the victims, and they had a higher likelihood of experiencing less damaging stings (OR 0.57; 95% CI 0.37 to 0.86). The TPIC was notified within 6 hours of the sting incidence in 356 (48.1%) cases that showed significant relationships with lower severity (OR 3.51; 95% CI 1.9 to 6.3). However, only 15% of the scorpions could be identified (Marks et al. 2019).

4.4   SOME COMMON CLINICAL SYMPTOMS OF SCORPION ENVENOMATION

The following three levels of severity of scorpion envenomation have been identified (Goyffon et al. 1982).

Stage I is associated with a low venom dose, resulting in non-lethal envenomations. Pain is quick, acute, and continuous, with sensations of partial recoveries and relapses. It lasts from the 15th to the 24th hour and occasionally even longer. The only clinical symptom is usually discomfort, which can last up to 2 hours after being stung. This pain can occur alone or in conjunction with other general symptoms such as anxiety, fever, sweats, nausea, a general feeling of faintness, and an alternation of high and low blood pressure—the transition to the next stage is frequently sudden and without notice. Complications are likely between the 3rd and 12th hours, considered critical.

Severe envenomation belongs to Stage II. Systemic muscarinic symptoms appear within 2–3 hours of the sting as sweating, vomiting, epigastric pain, diarrhea, colics, priapism, hypotension, pulmonary obstruction, bradycardia, dyspnea (shortness of breath), and patients feel intense heat. Vomiting is a severe symptom and requires close monitoring; in 5–10% of Stage II cases, entry into the next, harsh stage occurs between the 4th and the 12th hours after the sting.

Stage III reflects extremely severe poisoning, potentially fatal. Cardiovascular collapse and serious respiratory consequences such as pulmonary edema, bronchospasm, and cyanosis are life-threatening (Gueron and Ovsyshcher 1987; Gueron and Yaron 1970). Symptoms of myocardial ischemia, such as hyperthermia, cardiac arrhythmias, and electrocardiographic signs, may be seen. At this point, nearly half of the cases will experience a catastrophic progression within a few hours or perhaps minutes (Nouira et al. 2007).

Some enzyme inhibitors, proteins known as neurotoxic peptides, mucopolysaccharides, hyaluronidase, phospholipase A_2, serotonin, histamine, and others are found in scorpion venom (Quintero-Hernández et al. 2011; Petricevich 2010; Das et al. 2020; Pessini et al. 2001). Scorpion venom may contain neurotoxin, cardiotoxin, nephrotoxin, hemolytic toxin, etc. Voltage-dependent ion channels are the main targets of scorpion venom. By interfering with ion channels, the neurotoxic peptides (neurotoxins) in the venom produce the symptoms of envenomation (Mouhat et al. 2004). The long-chain polypeptide neurotoxin stabilizes voltage-dependent sodium channels in the open state, which results in continuous, prolonged, repetitive firing of somatic, sympathetic, and parasympathetic neurons. This overexcitation of the autonomic and neuromuscular systems causes the release of excessive amounts of neurotransmitters such as epinephrine, nor-epinephrine, acetylcholine, glutamate, and aspartate, which have the potential to cause several clinical symptoms (mentioned earlier) (Natu et al. 2006).

Ion channels are gated pores, and binding or changes in the voltage gradient can either intrinsically gate or control the channel. This pattern is influenced by hormone production, nerve and muscle stimulation, cell proliferation, sensory transduction, salt and water balance regulation, and blood pressure (Ashcroft and Gribble 2000). Due to their exceptional affinity and selectivity, scorpion toxins have been employed as pharmacological instruments to characterize an array of receptor proteins implicated in regular ion channel activity and aberrant channel function under pathological conditions (Lecomte et al. 1998; Lehmann-Horn and Jurkat-Rott 1999).

Depending on the type of scorpion, the sympathetic and parasympathetic nervous systems are primarily responsible for mediating the clinical signs of envenomation. It is well known that neurotoxins in scorpion venom activate calcium channels in pre-synaptic nerve terminals, inducing autonomic storms (Santos et al. 2016). Because the excitatory neurotoxins in scorpion venom cause significant autonomic neuro-transmitter release (adrenergic or cholinergic autonomic storm), the systemic effects of scorpion envenomation typically include the release of catecholamines. They release the catecholamines from the adrenal medulla (directly and through parasym-pathetic stimulation), stimulating peripheral sympathetic nerve terminals and making the venom a potent arrhythmogenic agent. It sets off a series of events that may lead to mortality, unconsciousness, tachycardia, bradycardia, arrhythmia, arterial hypo-tension or hypertension, pulmonary edema, and heart failure (Isbister and Bawaskar 2014; Bawaskar and Bawaskar 1996).

Furthermore, clinical accounts of scorpion sting patients showed cardiac dys-function and respiratory failure, which might be lethal (Cesaretli and Ozkan 2010; Gazarian et al. 2005; Oukkache et al. 2008; Sami-Merah et al. 2008; Abroug et al. 1999). Immune system sensitivity and the relationship between venom dosage and patient body weight have both been linked to high rates of morbidity and death in youngsters (Dabo et al. 2011; Goyffon and Billiald 2007). Scorpion stings in chil-dren under seven are considered dangerous because their immune systems are still forming or failing; however, healthy adults are not immune to scorpion stings. This category also includes fatal events (Matthiensen 1988; Soares and Azevedo 2002). In addition to the patient's age and sensitivity, the type of scorpion, the patient's sex, the site of the sting, and the amount of time that passes between the moment of the sting and the need for first aid are other known variables that may affect the symptoms of a scorpion bite. Because the patient may experience a possibly systemic crisis within minutes or hours after a scorpion sting, it is crucial to respond to the sting promptly, which may involve bringing the patient to the closest hospital (de Roodt et al. 2003; Porto 2010; Osnaya-Romero et al. 2001).

When a scorpion envenomates a person, they may experience intense local pain, which may or may not be associated with tissue damage. Only a tiny percentage of scorpion envenomation causes systemic consequences. The effects vary depending on the scorpion species and are induced by various excitatory neurotoxins (Uluğ et al. 2012; Isbister et al. 2003). The percentage of scorpion sting involvement in the lower limb ranged from 15% to 60.4%, whereas upper limb involvement ranged from 12.3% to 73.7%. The percentages in the trunk varied from 3.8% to 17.3%, and 0.9% to 8.6% in the head and neck (Adiguzel et al. 2007; Bahloul et al. 2010; de Roodt et al. 2003; Gómez et al. 2010; Mohseni et al. 2013; Pardal et al. 2003; Shahbazzadeh et al. 2009; Bosnak et al. 2009).

According to Petrecevich (2004), scorpion venom stimulates inflammatory cells, platelet-activating factors, leukocytes, immunoglobulins, adhesion molecules, and cytokines linked to immune system dysfunctions. A cascade of cell components, systems, and mediator release sets off the inflammatory response (Fabiano et al. 2008). Numerous studies substantiate the involvement of cytokines in scorpion envenomation; pro- and anti-inflammatory cytokine levels appear to be elevated in sepsis syndrome. They still lack clarity about their clinical importance and prognostic value (Petricevich 2004, 2006; Petricevich and Pena 2002; Sofer 1995; Magalhães et al. 1999; Fukuhara et al. 2003).

As mentioned earlier, the envenomation process produces cytokines in a cascade (Table 4.1). Tissue damage has a complicated etiology, making it impossible to pinpoint a single cause. Inflammation causes tissue damage, which can escalate and ultimately result in organ failure due to malfunction. For damaged tissue to heal structurally and functionally, cytokines must be divided into pro- and anti-inflammatory responses. However, because inflammatory cells produce products, the overproduction of proinflammatory signals can worsen tissue injury. Bradykinin, prostaglandins, catecholamines, and corticosteroids are among the substances scorpion venoms release that can activate the neuroendocrinal–immunological axis. All these agents proved to induce the release of immunological mediator cytokines. According to reports, the cytotoxin from *H. lepturus* venom can lead to deadly

---

TABLE 4.1
Inflammatory mediators are involved in scorpion venom-treated animal and human serum

Scorpion Species	Pre-clinical Study (Model) /Human Clinical Data	Cytokines Produced	References
Androctonus australis hector	Rats	IL-1β, IL-4, IL-6, IL-10, and TNF-α.	Adi-Bessalem et al. 2008
Buthus martensi Karch	Rat	NO and paw oedema	Liu et al. 2008
Centruroides noxius	Mice	IL-1β, IL-1α, IFN-γ IL-6, IL-10, and TNF-α.	Petricevich 2006
Hemiscorpius lepturus	Experiment with human monocytes and venom-induced human serum	IL-12, TNF-α.	Hadaddezfuli et al. 2015; Jalali et al. 2011
Leiurus quinquestriatus	Clinical data, Rabbit (animal)	IL-6, IL-8, NO, and TNF-α.	Sofer 1995; Abdoon and Fatani 2009
Tityus serrulatus	Clinical data, Rabbit (animal)	IL-1β, IL-6, IL-8, IL-10, IFN-γ, IL-1α, -1β, NO, TNF-α, and GM-CSF.	; Fukuhara et al. 2003; Petricevich 2002; Petricevich et al. 2007; Petricevich et al. 2008

hemolysis, hematuria, ankylosis of the joints, hemoglobinuria, cutaneous necrosis, anxiety, depression, mental disorders, schizophrenia, and even death (Pipelzadeh et al. 2006; Rahmani and Jalali 2012; Radmanesh 1990; Shahi et al. 2015).

Pulmonary edema is the most severe respiratory consequence that scorpion-stung patients experience. While some clinical findings link pulmonary edema to increased capillary permeability, most clinical and experimental investigations point to a hemo-dynamic basis for this disease (Bahloul et al. 2013; Sofer and Gueron, 1988). Sudden pulmonary edema has been related to myocardial injury from venom or premature left ventricular failure caused by a significant catecholamine release (Bahloul et al. 2006). There has been substantial debate in the past over the source of ventricular dysfunction and myocardial damage brought on by severe scorpion envenomation (Bahloul et al. 2013).

The initial concept that gained acceptance suggested that the elevated catecholamine levels were caused by the venom's direct stimulation of the adrenal gland and sym-pathetic nervous system. The second acknowledged theory is the venom's immediate impact on the heart, which results in "scorpion myocarditis" (Bahloul et al. 2013). It is likely that the venom directly affects the cardiac cell membranes, changing their perme-ability and electrical characteristics, leading to functional impairments through aberrant electrolyte fluxes and shifts. The final recognized theory is that myocardial ischemia may cause cardiac dysfunction. The last recognized approach is that myocardial ischemia may cause cardiac dysfunction (Mishra and Prasad 2015; Chippaux 2012).

Assessment of the severity of poisoning is essential to determine prognosis and implement effective treatment, especially in children (Santhanakrishnan and Raju 1974). The best way to treat scorpion stings is still a matter of debate. According to some experts (Abroug et al. 1999; Bawaskar and Bawaskar 2000; Das et al. 2020, 2022), ASA, the only specific treatment, is debatable. In contrast, symptomatic or supportive treatments, now universally acknowledged by experts, frequently depend on specific useful drugs, indications, and dosages (Freire-Maia et al. 1994). There have been attempts to define the prevalence and severity of scorpion stings world-wide and improve the management and treatment guidelines based on contemporary literature, discussed in Chapter 5.

4.5   CONCLUSION

Only a broad guide can be given to the number of species listed here, projected sting rates, geographic ranges, and sting classification. Scorpions are becoming increas-ingly diverse, with new species being discovered regularly. In the future, precise scorpion identification needs to be the main focus of any research on scorpions, par-ticularly in investigations assessing scorpionism and venom characterization to create novel medications. It is during the summer when people are most vulnerable to scor-pion stings. Therefore, the risk of scorpion stings can be considerably decreased by assembling a professional reinforcement team and making fewer summertime trips to high-risk locations. Furthermore, as scorpion stings are more common in rural regions, teaching rural residents in public settings like schools and among influential individuals can significantly lower the frequency and risk of sting incidences.

REFERENCES

Abdoon NA, and Fatani AJ. 2009. Correlation between blood pressure, cytokines and nitric oxide in conscious rabbits injected with *Leiurus quinquestriatus quinquestriatus* scorpion venom. *Toxicon* 54 (4):471–480.

Abroug F, ElAtrous S, Nouria S, Haguiga H, and Touzi N. 1999. Bouchoucha S. Serotherapy in scorpion envenomation: a randomised controlled trial. *Lancet* 354 (9182):906–909.

Abroug F, Nouira S, El Atrous S, Besbes L, Boukef R, Boussarsar M, Marghli S, Eurin J, Barthelemy C, El Ayeb M, and Dellagi K. 2003. A canine study of immunotherapy in scorpion envenomation. *Intensive Care Medicine* 29:2266–2276.

Adi-Bessalem S, Hammoudi-Triki D, and Laraba-Djebari F. 2008. Pathophysiological effects of *Androctonus australis hector* scorpion venom: tissue damages and inflammatory response. *Experimental and Toxicologic Pathology* 60 (4–5):373–380.

Adiguzel S, and Ozkan O. 2007. Inceoglu B. Epidemiological and clinical characteristics of scorpionism in children in Sanliurfa, Turkey. *Toxicon* 49 (6):875–880.

Agrawal A, Kumar A, Consul S, and Yadav A. 2015. Scorpion bite, a sting to the heart!. *Indian Journal of Critical Care Medicine* 19:233–236.

Ahsan MM, Tahir HM, & Naqi JA, 2015. First report of scorpion envenomization in District Sargodha, Punjab, Pakistan. *Biologia (Pakistan)* 61 (2):279–285.

Albuquerque CM, Santana Neto Pde L, Amorim ML, and Pires SC. 2013. Pediatric epidemiological aspects of scorpionism and report on fatal cases from *Tityus stigmurus* stings (Scorpiones: Buthidae) in State of Pernambuco, Brazil. *Revista da Sociedade Brasileira de Medicina Tropical* 46:484–489.

Alhamoud MA, Al Fehaid MS, Alhamoud MA, Alzoayed MH, Alkhalifah AA, and Menezes RG. 2021. Scorpion stings in Saudi Arabia: an overview. *Acta Bio Medica: Atenei Parmensis* 92 (4):e2021273.

Almitai Y, Mines Y, Aker M, and Goitein K. 1985. Scorpion sting in children. A review of 51 cases. *Clincal Pediatrics* 24 (3):136–140.

Amaral CFS, Barbosa AJ, Leite VHR, Tafuri WL, and de Rezende NA. 1994. Scorpion sting-induced pulmonary oedema: evidence of increased alveolocapillary membrane permeability. *Toxicon* 32 (8):999–1003.

Amaral CFS, and Rezende NA. 1997. Both cardiogenic and non-cardiogenic factors are involved in the pathogenesis of pulmonary oedema after scorpion envenoming. *Toxicon* 35:997–998.

Amitai Y. 1998. Clinical manifestations and management of scorpion envenomation. *Public Health Reviews* 26 (3): 257–263.

Ashcroft FM, and Gribble FM. 2000. Tissue-specific effects of sulfonylureas: lessons from studies of cloned KATP channels. *Journal of Diabetes and Its Complications* 14 (4):192–196.

Azevedo JLS. 2006. A importância dos centros de informação e assistência toxicológica e sua contribuição na minimização dos agravos à saúde e ao meio ambiente no Brasil, Universidade de Brasília:247.

Badhe RV, Thomas AB, Harer SL, Deshpande AD, Salvi N, and Waghmare A. 2006. Intraspecific variation in protein pattern of red scorpion (*Mesobuthus tamulus*, coconsis, pocock) venoms from Western and Southern India. *Journal of Venomous Animals and Toxins Including Tropical Diseases* 12:612–619.

Bahloul M, Ben Hamida C, Chtourou K, Ksibi H, Dammak H, Kallel H, Chaari A, Chelly H, Guermazi F, Rekik N, and Bouaziz M. 2004. Evidence of myocardial ischaemia in severe scorpion envenomation: myocardial perfusion scintigraphy study. *Intensive Care Medicine* 30:461–467.

Bahloul M, Chaari A, Dammak H, Samet M, Chtara K, Chelly H, Hamida CB, Kallel H, and Bouaziz M. 2013. Pulmonary oedema following scorpion envenomation: mechanisms, clinical manifestations, diagnosis and treatment. *International Journal of Cardiology* 162 (2):86–91.

Bahloul M, Chaari AN, Kallel H, Khabir A, Ayadi A, Charfeddine H, Hergafi L, Chaari AD, Chelly HE, Hamida CB, and Rekik N. 2006. Neurogenic pulmonary oedema due to traumatic brain injury: evidence of cardiac dysfunction. *American Journal of Critical Care* 15 (5):462–470.

Bahloul M, Chaari A, Khlaf-Bouaziz N, Hergafi L, Ksibi H, Kallel H, Chaari A, Chelly H, Hamida CB, Rekik N, and Bouaziz M. 2005. Gastrointestinal manifestations in severe scorpion envenomation. *Gastroentérologie Clinique et Biologique* 29 (10):1001–1005.

Bahloul M, Chabchoub I, Chaari A, Chtara K, Kallel H, Dammak H, Ksibi H, Chelly H, Rekik N, Hamida CB, and Bouaziz M. 2010. Scorpion envenomation among children: clinical manifestations and outcome (analysis of 685 cases). *The American Journal of Tropical Medicine and Hygiene* 83 (5):1084.

Barbosa AD, Magalhães DFD, Silva JAD, Silva MX, Cardoso MDFEC, Meneses JNC, and Cunha MDCM. 2012. Epidemiological study of scorpion stings in Belo Horizonte, Minas Gerais state, Brazil, 2005–2009. *Cadernos de Saúde Pública* 28:1785–1789.

Baseer KA, and Naser MA. 2019. Predictors for mortality in children with scorpion envenomation admitted to pediatric intensive care unit, Qena Governorate, Egypt. *The American Journal of Tropical Medicine and Hygiene* 101 (4):941.

Bawaskar HS, and Bawaskar PH. 1992. Management of the cardiovascular manifestations of poisoning by the Indian red scorpion (*Mesobuthus tamulus*). British Heart Journal 68 (11):478.

Bawaskar HS, and Bawaskar PH. 1996. Severe envenoming by the Indian red scorpion *Mesobuthus tamulus*: the use of prazosin therapy. *QJM: An International Journal of Medicine* 89 (9): 701–704.

Bawaskar HS, and Bawaskar PH. 1998. Indian red scorpion envenoming. *The Indian Journal of Pediatrics* 65:383–391.

Bawaskar HS, and Bawaskar PH. 1999. Management of scorpion sting. *Heart* 82 (2):253–254.

Bawaskar HS, and Bawaskar PH. 2000. Prazosin therapy and scorpion envenomation. *Journal of Association of Physicians of India* 48 (12):1175–1180

Bawaskar HS, and Bawaskar PH. 2012. Scorpion sting: update. *Journal of Association of Physicians of India* 60:46–55.

Bawaskar HS, Bawaakar PH, and Bawaskar PH. 2014. Homocysteine: often neglected but common culprit of coronary heart diseases. *Journal of Cardiovascular Disease Research* 5 (3):40.

Bertani R, Bonini RK, Toda MM, Isa LS, Alvarez Figueiredo JV, dos Santos MR, and Ferraz SC. 2018. Alien scorpions in the Municipality of São Paulo, Brazil-evidence of successful establishment of *Tityus stigmurus* (Thorell, 1876) and first records of *Broteochactas parvulus* (Pocock, 1897) and Jaguajir rochae (Borelli, 1910*). BioInvasions Records* 7 (1):89–94.

Bond GR. 1992. Antivenin administration for *Centruroides* scorpion sting: risks and benefits. *Annals of Emergency Medicine* 21 (7):788–791.

Bosnak M, Ece A, Yolbas I, Bosnak V, Kaplan M,and Gurkan F. 2009. Scorpion sting envenomation in children in southeast Turkey. *Wilderness & Environmental Medicine* 20 (2):118–124.

Bouaziz M, Ben Hamida C, Chelly H, Rekik N, Jeddi HM. 1996. Envenimation Paris: Arnette. *L'envenimation scorpionique: étude épidémiologique, clinique et éléments de prognostic*:11–35.

Bucaretchi F, Baracat E, Nogueira RN, and Zambrone FAD. 1995. Severe scorpion envenomation in children: a comparison between *Tityus bahiensis* and *Tiyus serrulatus*. *Revista do Instituto de Medicina Tropical de São Paulo* 37:331–336.

Cesaretli Y, and Ozkan O. 2010. Scorpion stings in Turkey: epidemiological and clinical aspects between the years 1995 and 2004. *Revista do Instituto de Medicina Tropical de São Paulo* 52:215–220.

Chippaux JP. 2012. Emerging options for the management of scorpion stings. *Drug Design, Development and Therapy* 6:165–173.

Chippaux JP. 2015. Epidemiology of envenomations by terrestrial venomous animals in Brazil based on case reporting: from obvious facts to contingencies. *Journal of Venomous Animals and Toxins Including Tropical Diseases* 21:1–17.

Chippaux JP, and Goyffon M. 2008. Epidemiology of scorpionism: a global appraisal. *Acta Tropica* 107 (2):71–79.

Chowell G, Hyman JM, Díaz-Dueñas P, and Hengartner NW. 2005. Predicting scorpion sting incidence in an endemic region using climatological variables. *International Journal of Environmental Health Research* 15 (6):425–435.

Cupo P. 2015. Clinical update on scorpion envenoming. *Revista da Sociedade Brasileira de Medicina Tropical* 48:642–649.

Dabo A, Golou G, Traoré MS, Diarra N, Goyffon M, and Doumbo O. 2011. Scorpion envenoming in the North of Mali (West Africa): epidemiological, clinical and therapeutic aspects. *Toxicon* 58 (2):154–158.

Das B, Patra A, and Mukherjee AK. 2020. Correlation of venom toxinome composition of Indian red scorpion (*Mesobuthus tamulus*) with clinical manifestations of scorpion stings: failure of commercial antivenom to immune-recognize the abundance of low molecular mass toxins of this venom. *Journal of Proteome Research* 19 (4):1847–1856.

Das B, Patra A, Puzari U, Deb P, and Mukherjee AK. 2022. In vitro laboratory analyses of commercial anti-scorpion (*Mesobuthus tamulus*) antivenoms reveal their quality and safety but the prevalence of a low proportion of venom-specific antibodies. *Toxicon* 215:37–48.

Das B, Saviola AJ, and Mukherjee AK. 2021. Biochemical and proteomic characterization, and pharmacological insights of Indian red scorpion venom toxins. *Frontiers in Pharmacology* 12:710680.

de Roodt AR, García SI, Salomón OD, Segre L, Dolab JA, Funes RF, and de Titto EH. 2003. Epidemiological and clinical aspects of scorpionism by *Tityus trivittatus* in Argentina. *Toxicon* 41 (8):971–977.

de Sousa Alves R, do Nascimento NRF, Barbosa PSF, Kerntopf MR, Lessa LMA, De Sousa CM, Martins RD, Sousa DF, De Queiroz MGR, Toyama MH and Fonteles MC. 2005. Renal effects and vascular reactivity induced by *Tityus serrulatus* venom. *Toxicon* 46 (3):271–276.

Dehesa-Dávila M. 1989. Epidemiological characteristics of scorpion sting in Leon, Guanajuato, Mexico. *Toxicon* 27 (3):281–286.

Dehesa-Dávila M, and Possani LD. 1994. Scorpionism and serotherapy in Mexico. *Toxicon* 32 (9):1015–1018.

Dehghani R, Doroudgar A, Khademi MR, Sayyah M. 1998. The survey of scorpion sting in the city of Kashan. *Journal of Esfahan University of Medical Sciences Health Services* 3 (2):132–135.

Dehghani R, and Fathi B. 2012. Scorpion sting in Iran: a review. *Toxicon* 60 (5):919–933.

Dehghani R, Vazirianzadeh B, Nasrabadi MR, and Moravvej SA. 2010. Study of scorpionism in Kashan in central of Iran. *Pakistan Journal of Medical Sciences* 26 (4):955–958.

Dutertre S, and Lewis RJ. 2010. Use of venom peptides to probe ion channel structure and function. *Journal of Biological Chemistry* 285 (18):13315–13320.

Ebrahimi V, Hamdami E, Moemenbellah-Fard MD, and Jahromi SE. 2017. Predictive determinants of scorpion stings in a tropical zone of south Iran: use of mixed seasonal autoregressive moving average model. *Journal of Venomous Animals and Toxins Including Tropical Diseases* 23:39.

El-Amin EO, Sultan OM, Al-Magamci MS, and Elidrissy A. 1994. Serotherapy in the management of scorpion sting in children in Saudi Arabia. *Annals of Tropical Paediatrics* 14 (1):21–24.

Emam SJ, Khosravi AD, and Alemohammad A. 2008. Evaluation of hematological and urine parameters in Hemiscorpius lepturus (Gadim) victims referred to Razi hospital, Ahwaz, Iran. *Journal of Medical Sciences* 8:306–309 (in Persian, abstract in English).

Fabiano G, Pezzolla A, Filograna MA, and Ferrarese F. 2008. Traumatic shock—physiopathologic aspects. *Il Giornale di chirurgia* 29 (1–2):51–57.

Freire-Maia L, Campos JA, and Amaral CF. 1994. Approaches to the treatment of scorpion envenoming. *Toxicon* 32 (9):1009–1014.

Freire-Maia L, Pinto GI, and Franco I. 1974. Mechanism of the cardiovascular effects produced by purified scorpion toxin in the rat. *Journal of Pharmacology and Experimental Therapeutics* 188 (1):207–213.

Fukuhara YDM, Reis MLD, Dellalibera-Joviliano R, Cunha FQC, and Donadi EA. 2003. Increased plasma levels of IL-1β, IL-6, IL-8, IL-10 and TNF-α in patients moderately or severely envenomed by *Tityus serrulatus* scorpion sting. *Toxicon* 41 (1):49–55.

Gazarian KG, Gazarian T, Hernández R, Possani LD. 2005. Immunology of scorpion toxins and perspectives for generation of antivenom vaccines. *Vaccine* 23 (26):3357–3368.

Gomes A, and Gomes A. 2015. Scorpion venom research around the world: Heterometrus species. *Scorpion Venoms: Springer* 351–367.

Gomes JV, Fé NF, Santos HL, Jung B, Bisneto PF, Sachett A, de Moura VM, da Silva IM, de Melo GC, de Oliveira Pardal PP, and Lacerda M. 2020. Clinical profile of confirmed scorpion stings in a referral center in Manaus, Western Brazilian Amazon. *Toxicon.* 1 (187):245–254.

Gómez JP, Quintana JC, Arbeláez P, Fernández J, Silva JF, Barona J, Gutiérrez JC, Díaz A, and Otero R. 2010. *Tityus asthenes* scorpion stings: epidemiological, clinical and toxicological aspects. *Biomedica* 30 (1):126–139.

Goyffon M, and Billiald, P. 2007. Envenomations VI. Scorpionism in Africa. *Medecine Tropicale: Revue du Corps de Sante Colonial* 67 (5):439–446.

Goyffon M, Vachon M, and Broglio N. 1982. Epidemiological and clinical characteristics of the scorpion envenomation in Tunisia. *Toxicon* 20 (1):337–344.

Guéguinou M, Chantôme A, Fromont G, Bougnoux P, Vandier C, and Potier-Cartereau M. 2014. KCa and Ca²⁺ channels: the complex thought. Biochimica et Biophysica Acta (BBA)-Molecular Cell Research 1843 (10):2322–2333.

Gueron M, and Ovsyshcher I. 1987. What is the treatment for the cardiovascular manifestations of scorpion envenomation?. *Toxicon: Official Journal of the International Society on Toxinology* 25:121–130.

Gueron M, and Yaron R. 1970. Cardiovascular manifestations of severe scorpion sting: clinicopathologic correlations. *Chest* 57 (2):156–162.

Guieu, R, Kopeyan C, Sampieri F, Devaux C, Bechis G, and Rochat H. 1995 Use of antrolene in experimental scorpion envenomation by *Androctonus australis Hector*. *Archives of Toxicology* 69 (8):575–577.

Guinand A, Cortés H, Díaz P, Sevcik C, González-Sponga M, and Eduarte G. 2004. Escorpionismo del género *Tityus* en la sierra falconiana y su correlación con la liberación de mediadores inflamatorios y enzimas cardíacas. *Gaceta Médica de Caracas* 112 (2):131–138.

Hadaddezfuli R, Khodadadi A, Assarehzadegan MA, Pipelzadeh MH, and Saadi S. 2015. *Hemiscorpius lepturus* venom induces expression and production of interluckin-12 in human monocytes. *Toxicon* 100:27–31.

Hauke TJ, and Herzig V. 2017. Dangerous arachnids—fake news or reality? *Toxicon* 138:173–183.

Isbister GK, and Bawaskar HS. 2014. Scorpion envenomation. *New England Journal of Medicine* 371 (5):457–463.

Isbister GK, Volschenk ES, Balit CR, and Harvey MS. 2003. Australian scorpion stings: a prospective study of definite stings. *Toxicon* 41 (7):877–883.

Jalali A, Pipelzadeh MH, Taraz M, Khodadadi A, Makvandi M, and Rowan EG. 2011. Serum TNF-α levels reflect the clinical severity of envenomation following a Hemiscorpius lepturus sting. *European Cytokine Network* 22(1):5–10.

Jalali A, and Rahim F. 2014. Epidemiological review of scorpion envenomation in Iran. *Iranian Journal of Pharmaceutical Research: IJPR* 13 (3):743.

Jarrar BM, and Al-Rowaily MA. 2008. Epide- miological aspects of scorpion stings in Al-Jouf Province, Saudi Arabia. *Annals of Saudi Medicine* 28:183–187.

Kankonkar RC, Kulkurni DG, and Hulikavi CB. 1998. Preparation of a potent anti-scorpion-venom-serum against the venom of red scorpion (*Buthus tamalus*). *Journal of Postgraduate Medicine* 44 (4):85–93.

Khan MF, and Ullah H. 2017. Multi-orgasn dysfunction secondary to yellow scorpion sting. *Journal of Ayub Medical College* 29 (2): 347–349.

Khatony A, Abdi A, Fatahpour T, and Towhidi F. 2015. The epidemiology of scorpion stings in tropical areas of Kermanshah province, Iran, during 2008 and 2009. *Journal of Venomous Animals and Toxins Including Tropical Diseases* 21. DOI: https://doi.org/10.1186/s40409-015-0045-4.

Krifi MN, Kharrat H, Zghal K, Abdouli M, Abroug F, Bouchoucha S, Dellagi K, and El Ayeb M. 1998. Development of an ELISA for the detection of scorpion venoms in sera of humans envenomed by *Androctonus australis garzonii* (Aag) and *Buthus occitanus tunetanus* (Bot): correlation with clinical severity of envenoming in Tunisia. *Toxicon* 36 (6):887–900.

Kularatne SA, Dinamithra NP, Sivansuthan S, Weerakoon KG, Thillaimpalam B, Kalyanasundram V, and Ranawana KB. 2015. Clinico-epidemiology of stings and envenoming of *Hottentotta tamulus* (Scorpiones: Buthidae), the Indian red scorpion from Jaffna Peninsula in northern Sri Lanka. *Toxicon* 93:85–89.

Laustsen AH, Solà M, Jappe EC, Oscoz S, Lauridsen LP, and Engmark M. 2016. Biotechnological trends in spider and scorpion antivenom development. *Toxins* 8 (8):226.

Lecomte C, Sabatier JM, Van Rietschoten J, and Rochat H. 1998. Synthetic peptides as tools to investigate the structure and pharmacology of potassium channel-acting short-chain scorpion toxins. *Biochimie* 80 (2):151–154.

Lehmann-Horn F, and Jurkat-Rott K. 1999. Voltage-gated ion channels and hereditary disease. *Physiological Reviews* 79 (4):1317–1372.

Liu T, Pang XY, Jiang F, and Ji YH. 2008. Involvement of spinal nitric oxide (NO) in rat pain-related behaviors induced by the venom of scorpion Buthus martensi Karsch. *Toxicon* 52 (1):62–71.

Lourenço WR. 2018. The evolution and distribution of noxious species of scorpions (Arachnida: Scorpiones). *Journal of Venomous Animals and Toxins Including Tropical Diseases* 24, 1.

Lourenço WR, Cloudsley-Thompson JL, Cuellar O, Eickstedt VV, Barraviera B, and Knox MB. 1996. The evolution of scorpionism in Brazil in recent years. *Journal of Venomous Animals and Toxins* 2:121–134.

Madhavan J. 2015. *A Study on Clinical Presentation and Outcome of Scorpion Sting in Children.* Thanjavur: Doctoral dissertation, Thanjavur Medical College.

Magalhães MM, Pereira MES, Amaral CF, Rezende NA, Campolina D, Bucaretchi F, Gazzinelli RT, and Cunha-Melo JR. 1999. Serum levels of cytokines in patients envenomed by *Tityus serrulatus* scorpion sting. *Toxicon* 37 (8):1155–1164.

Maghsoodi N, Vazirianzadeh B, and Salahshoor A. 2015. Scorpion sting in Izeh, Iran: an epidemiological study during 2009–2011. *Journal of Basic & Applied Sciences* 11:403–409.

Mahadevan, S. 2000. Scorpion sting. *Indian Pediatrics* 37:504–514.

Marks CJ, Muller GJ, Sachno D, Reuter H, Wium CA, Du Plessis CE, and Van Hoving DJ. 2019. The epidemiology and severity of scorpion envenoming in South Africa as managed by the Tygerberg Poisons Information Centre over a 10 year period. *African Journal of Emergency Medicine* 9 (1):21–24.

Matthiensen FA. 1988. Os escorpiões e suas relações com o homem: uma revisão. *Ciência e Cultura* 40 (12):1168–1172.

Ministério da Saúde (MS). 2009. Secretaria de Vigilância em Saúde (SVS). Manual de Controle de Escorpiões. Brasília: MS:74.

Ministério da Saúde (MS). 2018. Secretaria de Vigilância em Saúde (SVS). Sistema de Informação de Agravos de Notificação (SINAN). Saúde de A a Z. Acidentes-por-animais-peconhentos. Brasília: MS; 2018.

Mishra OP, and Prasad R. 2015. Myocardial dysfunction in children with scorpion sting envenomation. *Indian Pediatrics* 52 (4):291–292.

Mohseni A, Vazirianzadeh B, Hossienzadeh M, Salehcheh M, Moradi A, and Moravvej SA. 2013. The roles of some scorpions, *Hemiscorpius lepturus* and *Androctonus crassicauda*, in a scorpionism focus in Ramhormorz, southwestern Iran. *Journal of Insect Science* 13 (1):89.

More SS, Kiran KM, and Gadag JR. 2004. Dose-dependent serum biochemical alterations in Wistar albino rats after *Palamneus gravimanus* (Indian black scorpion) envenomation. *Journal of Basic and Clinical Physiology and Pharmacology* 15 (3–4):263–276.

Mullen G, and Durden L. 2019. Medical and veterinary entomology. *Annals of Tropical Medicine and Parasitology.* Vol. 3. Cambridge, US: Academic Press.

Murthy KR, and Zare MA. 1998. Effect of Indian red scorpion (*Mesobuthus tamulus concanesis, Pocock*) venom on thyroxine and triiodothyronine in experimental acute myocarditis and its reversal by species specific antivenom. *Indian Journal of Experimental Biology* 36 (1):16–21.

Natu VS, Murthy RKK, and Deodhar KP. 2006. Efficacy of species-specific anti-scorpion venom serum (AScVS) against severe serious scorpion stings (*Mesobuthus tamulus concanesis Pocock*)—an experience from rural hospital in Western Maharashtra. *Journal of the Association of Physicians of India* 54:283–287.

Navidpour S, Kovařík F, Soleglad ME, and Fet V. 2008. Scorpions of Iran (arachnida, scorpiones). Part i. Khoozestan province. *Euscorpius* 2008 (65):1–41.

Newton KA, Clench MR, Deshmukh R, Jeyaseelan K, and Strong PN. 2007. Mass fingerprinting of toxic fractions from the venom of the Indian red scorpion, *Mesobuthus tamulus*: biotope-specific variation in the expression of venom peptides. *Rapid*

Communications in Mass Spectrometry: An International Journal Devoted to the Rapid Dissemination of Up-to-the-Minute Research in Mass Spectrometry 21 (21):3467–3476.

Nouira S, Boukef R, Nciri N, Haguiga H, Elatrous S, Besbes L, Letaief M, and Abroug F. 2007. A clinical score predicting the need for hospitalization in scorpion envenomation. *The American Journal of Emergency Medicine* 25 (4):414–419.

Ortiz E, Gurrola GB, Schwartz EF, and Possani LD. 2015. Scorpion venom components as potential candidates for drug development. *Toxicon* 93:125–135.

Osnaya-Romero N, de Jesus Medina-Hernández T, Flores-Hernández SS, and León-Rojas G. 2001. Clinical symptoms observed in children envenomated by scorpion stings, at the children's hospital from the State of Morelos, Mexico. *Toxicon* 39 (6):781–785.

Oswald GA, Smith CC, Delamothe AP, Betteridge DJ, and Yudkin JS. 1988. Raised concentrations of glucose and adrenaline and increased in vivo platelet activation after myocardial infarction. *British Heart Journal* 59 (6):663.

Otero R, Navío E, Céspedes FA, Núñez MJ, Lozano L, Moscoso ER, Matallana C, Arsuza NB, García J, Fernández D, and Rodas, JH. 2004. Scorpion envenoming in two regions of Colombia: clinical, epidemiological and therapeutic aspects. *Transactions of the Royal Society of Tropical Medicine and Hygiene* 98 (12):742–750.

Oukkache N, Rosso JP, Alami M, Ghalim N, Saïle R, Hassar M, Bougis PE, and Martin-Eauclaire MF. 2008. New analysis of the toxic compounds from the *Androctonus mauretanicus mauretanicus* scorpion venom. *Toxicon* 51 (5):835–852.

Özkan Ö, and Karaer KZ. 2003. The scorpions in Turkey. *Turkish Bulletin of Hygiene and Experimental Biology* 60:55–62.

Ozkan O, and Kat I. 2005. Mesobuthus eupeus scorpionism in Sanliurfa region of Turkey. *Journal of Venomous Animals and Toxins including Tropical Diseases* 11:479–491.

Pardal PP, Castro LC, Jennings E, Pardal JS, and Monteiro MR. 2003. Epidemiological and clinical aspects of scorpion envenomation in the region of Santarém, Pará, Brazil [in Portuguese *Revista da Sociedade Brasileira de Medicina Tropical* 36 (3):349–353.

Pardal PPO, Cardoso BS, and Fan HW. 1999. Escorpionismo na região do rio Tapajós, Itaituba (Pará). *Revista da Sociedade Brasileira de Medicina Tropical* 32:394.

Peres ACP, Nonaka PN, de Carvalho PDTC, Toyama MH, Melo CA, de Paula Vieira R, Dolhnikoff M, Zamuner SR, and de Oliveira LVF. 2009. Effects of *Tityus serrulatus* scorpion venom on lung mechanics and inflammation in mice. *Toxicon* 53 (7–8):779–785.

Pessini AC, Kanashiro A, Malvar DDC, Machado RR, Soares DM, Figueiredo MJ, Kalapothakis E, and Souza GE. 2008. Inflammatory mediators involved in the nociceptive and oedematogenic responses induced by *Tityus serrulatus* scorpion venom injected into rat paws. *Toxicon* 52 (7):729–736.

Pessini AC, Takao TT, Cavalheiro EC, Vichnewski W, Sampaio SV, Giglio JR, and Arantes EC. 2001. A hyaluronidase from *Tityus serrulatus* scorpion venom: isolation, characterization and inhibition by flavonoids. *Toxicon* 39 (10):1495–1504.

Petricevich VL. 2002. Effect of *Tityus serrulatus* venom on cytokine production and the activity of murine macrophages. *Mediators of Inflammation* 11:23–31.

Petricevich VL. 2004. Cytokine and nitric oxide production following severe envenomation. *Current Drug Targets-Inflammation & Allergy* 3 (3):325–332.

Petricevich VL. 2006. Balance between pro-and anti-inflammatory cytokines in mice treated with *Centruroides noxius* scorpion venom. *Mediators of Inflammation* 2006 (6):54273.

Petricevich VL. 2010. Scorpion venom and the inflammatory response. *Mediators of Inflammation* 2010:903295.

Petricevich VL, Cruz AH, Coronas FI, and Possani LD. 2007. Toxin gamma from *Tityus serrulatus* scorpion venom plays an essential role in immunomodulation of macrophages. *Toxicon* 50(5):666–675.

Petricevich VL, and Pena CF. 2002. The dynamics of cytokine and nitric oxide secretion in mice injected with *Tityus serrulatus* scorpion venom. *Mediators of Inflammation* 11:173–180.

Petricevich VL, Reynaud E, Cruz AH, and Possani LD. 2008. Macrophage activation, phago-cytosis and intracellular calcium oscillations induced by scorpion toxins from *Tityus serrulatus*. *Clinical & Experimental Immunology* 154 (3):415–423.

Porto TJ, Brazil TK, and Lira-da-Silva RM. 2010. Scorpions, state of Bahia, northeastern Brazil. *Check List* 6 (2):292–297.

Possani LD, Merino E, Corona M, Bolivar F, and Becerril B. 2000. Peptides and genes coding for scorpion toxins that affect ion-channels. *Biochimie* 82 (9–10):861–868.

Quintero-Hernández V, Jiménez-Vargas JM, Gurrola GB, Valdivia HH, and Possani L. 2013. Scorpion venom components that affect ion-channels function. *Toxicon* 76:328–342.

Quintero-Hernández V, Ortiz E, Rendón-Anaya M, Schwartz EF, Becerril B, Corzo G, and Possani LD. 2011. Scorpion and spider venom peptides: gene cloning and peptide expression. *Toxicon* 58 (8):644–663.

Radmanesh M. 1990. Clinical study of *Hemiscorpion lepturus* in Iran. *The Journal of Tropical Medicine and Hygiene* 93:327–332.

Rafeeazadeh S. 2009. Report of scorpion sting In Iran during 2009. *Center of Management of Preventing and Fighting with the Diseases* 203:1–15.

Rahmani AH, and Jalali A. 2012. Symptom patterns in adult patients stung by scorpions with emphasis on coagulopathy and hemoglubinuria. *Journal of Venomous Animals and Toxins Including Tropical Diseases* 18:427–431.

Rajarajeswari G, Sivaprakasam S, and Viswanathan J. 1979. Morbidity and mortality pattern in scorpion stings. (A review of 68 cases). *Journal of the Indian Medical Association* 73 (7–8):123–126.

Rajashekhar MS. 2017. Epidemiological and clinical study of scorpion envenomation in patients admitted at Rims teaching hospital, Raichur. *International Journal of Scientific Study* 5 (3):73–76.

Ratnayake RM, Kumanan T, and Selvaratnam G. 2016. Acute myocardial injury after scorpion (*Hottentotta tamulus*) sting. *Journal of Ceylon Medicine* 61:86–87.

Rao VR, Perez-Neut M, Kaja S, and Gentile S. 2015. Voltage-gated ion channels in cancer cell proliferation. *Cancers* 7 (2):849–875.

Reckziegel GC, and Pinto VL. 2014. Scorpionism in Brazil in the years 2000 to 2012. *Journal of Venomous Animals and Toxins Including Tropical Diseases* 20:2–8.

Reddy R, Bompelli N, Khardenavis V, and Deshpande A. 2017. Scorpion bite-induced ischaemic stroke. *BMJ Case Reports* 2017:1.

Romero-Gutierrez T, Peguero-Sanchez E, Cevallos MA, Batista CV., Ortiz E, and Possani L D. 2017. A deeper examination of Thorellius atrox scorpion venom components with omic techonologies. *Toxins* 9 (12):399.

Sami-Merah S, Hammoudi-Triki D, Martin-Eauclaire MF, and Laraba-Djebari F. 2008. Combination of two antibody fragments F (ab')$_2$/Fab: an alternative for scorpion envenoming treatment. *International Immunopharmacology* 8 (10):1386–1394.

Santhanakrishnan BR, and Raju VB. 1974. Management of scorpion sting in children. *Journal of Tropical Medicine and Hygiene* 77 (6):133–135.

Santos MS, Silva CG, Neto BS, Grangeiro Junior CR, Lopes VH, Teixeira Junior AG, Bezerra DA, Luna JV, Cordeiro JB, Júnior JG, and Lima MA. 2016. Clinical and epidemiological aspects of scorpionism in the world: a systematic review. *Wilderness & Environmental Medicine* 27(4):504–518.

Sedziwy L, Thomas M, and Shillingford J. 1968. Some observations on haematocrit changes in patients with acute myocardial infarction. *British Heart Journal* 30 (3):344.

Severino DN, Pereira RL, Knysak I, Cândido DM, and Kwasniewski FH. 2009. Oedematogenic activity of scorpion venoms from the Buthidae family and the role of platelet-activating factor and nitric oxide in paw oedema induced by *Tityus* venoms. *Inflammation* 32:57–64.

Shahbazzadeh D, Amirkhani A, Djadid ND, Bigdeli S, Akbari A, Ahari H, Amini H. and Dehghani R. 2009. Epidemiological and clinical survey of scorpionism in Khuzestan province, Iran, 2003. *Toxicon* 53 (4):454–459.

Shahi M, Rafinejad J, Az-Khosravi L, and Moosavy SH. 2015. First report of death due to *Hemiscorpius acanthocercus* envenomation in Iran: Case report. *Electronic Physician* 7 (5):1234.

Sharma PP, Kaluziak ST, Pérez-Porro AR, González VL, Hormiga G, Wheeler WC, and Giribet G. 2014. Phylogenomic interrogation of Arachnida reveals systemic conflicts in phylogenetic signal. *Molecular Biology and Evolution* 31 (11):2963–2984.

Simms BA, and Zamponi GW. 2014. Neuronal voltage-gated calcium channels: structure, function, and dysfunction. *Neuron* 82 (1):24–45.

Soares MR, de Azevedo CS, and De MM. 2002. Scorpionism in Belo Horizonte, MG: a retrospective study. *Revista da Sociedade Brasileira de Medicina Tropical* 35(4):359–363.

Soares MRM, de Azevedo CS, and De MM. 2002. Scorpionism in Belo Horizonte, MG: a retrospective study. *Revista da Sociedade Brasileira de Medicina Tropical* 35 (4):359–63.

Sofer S. 1995. Scorpion envenomation. *Intensive Care Medicine* 21:626–628.

Sofer S, and Gueron M. 1988. Respiratory failure in children following envenomation by the scorpion *Leiurus quinquestriatus*: hemodynamic and neurological aspects. *Toxicon* 26 (10):931–939.

Sofer S, and Gueron M. 1990. Vasodilators and hypertensive encephalopathy following scorpion envenomation in children. *Chest* 97 (1):118–26.

Soulaymani-Bencheikh R, Faraj Z, Semlali I, Khattabi A, Skalli S, Benkirane R, and Badri M. 2002. Epidemiological aspects of scorpion stings in Morocco. *Revue d'epidemiologie et de Sante Publique* 50:341–347.

Soulaymani-Bencheikh R, Khattabi A, Faraj Z, and Semlali I. 2008. Management of scorpion sting in Morocco [in French]. *Annales Francaises* 27:317–322.

Srinivasan KN, Gopalakrishnakone P, Tan PT, Chew KC, Cheng B, Kini RM, Koh JL, Seah SH, and Brusic V. 2002. SCORPION, a molecular database of scorpion toxins. *Toxicon* 40 (1):23–31.

Suranse V, Sawant NS, Bastawade DB, and Dahanukar N. 2019. Haplotype diversity in medically important red scorpion (Scorpiones: Buthidae: *Hottentotta tamulus*) from India. *Journal of Genetics* 98 (1):17.

Suranse V, Sawant NS, Paripatyadar SV, Krutha K, Paingankar MS, Padhye AD, Bastawade DB, and Dahanukar N. 2017. First molecular phylogeny of scorpions of the family Buthidae from India. *Mitochondrial DNA Part A*. 28 (4):606–611.

Tiwari AK, and Deshpande SB. 1993. Toxicity of scorpion (*Buthus tamulus*) venom in mammals is influenced by the age and species. *Toxicon* 31 (12):1619–1622.

Torrez PP, Quiroga MM, Abati PA, Mascheretti M, Costa,WS, Campos LP, and França FO. 2015. Acute cerebellar dysfunction with neuromuscular manifestations after scorpionism presumably caused by *Tityus obscurus* in Santarém, Pará/Brazil. *Toxicon* 96:68–73.

Uluğ M, Yaman Y, Yapici F, and Can-Uluğ N. 2012. Scorpion envenomation in children: an analysis of 99 cases. *The Turkish Journal of Pediatrics* 54 (2):119–127.

Vaucel JA, Larréché S, Paradis C, Courtois A, Pujo JM, Elenga N, Résière D, Caré W, de Haro L, Gallart JC, and Torrents R. 2022. French Scorpionism (mainland and oversea territories): narrative review of scorpion species, scorpion venom, and envenoming management. *Toxins* 14 (10):719.

Vazirianzadeh BA, Farhadpour F, Hosseinzadeh M, Zarean M, and Moravvej SA. 2012. An epidemiological and clinical study on scorpionism in hospitalized children in Khuzestan, Iran. *Journal of Arthropod-Borne Diseases* 6(1):62.

Von Eickstedt VRD, Ribeiro LA, Candido DM, Albuquerque MJ, and Jorge MT. 1996. Evolution of scorpionism by (Perty) and *Tityus bahiensis* Tityus serrulatus Lutz and Mello and geographical distribution of the two species in the state of São Paulo-Brazil. *Journal of Venomous Animals and Toxins* 2:92–105.

Ward MJ, Ellsworth SA, and Nystrom GS. 2018. A global accounting of medically significant scorpions: epidemiology, major toxins, and comparative resources in harmless counterparts. *Toxicon* 151:137–155.

Yuvaraja K, Chidambaram N, Umarani R, Bhargav KM, Kumar SP, Prabhu T, et al. 2019. A study on clinical features, complications and management of scorpion sting envenomation at a tertiary care hospital, in rural South India. *Journal of Clinical and Scientific Research* 8:140.

Zhang L, Shi W, Zeng XC, Ge F, Yang M, Nie Y, Bao A, Wu S, and Guoji E. 2015. Unique diversity of the venom peptides from the scorpion Androctonus bicolor revealed by transcriptomic and proteomic analysis. *Journal of Proteomics* 14 (128):231–250.

5 Scorpion Sting Envenomation

Prevention and Treatment

5.1 ANTISCORPION ANTIVENOM THERAPY FOR HOSPITAL MANAGEMENT OF SCORPION STING

In many places, scorpion stings frequently result in emergencies (Chippaux and Goyffon 2008). Enhancing the handling of scorpion stings has reignited the debate on how best to treat them. Passive immunotherapy versus symptomatic treatment is the subject of disagreement that mainly emerged in Africa and Asia. This contradiction is partly due to the development of medical intensive care services, which spread to provincial hospitals in several scorpion-affected nations. Consequently, accessibility and symptomatic treatment advancements made the latter option a viable emergent treatment option. Only a tiny percentage of professionals are exclusively involved in medical acute care. ASAs must be of the most outstanding caliber for administration, both in terms of safety and efficacy, and they must be given as soon as possible to eliminate the venom for maximum effectiveness. Because most modern ASAs are composed of highly pure immunoglobulin G (IgG) fragments, their administration is made more accessible even in outlying clinics without critical care units. They are safer without sacrificing their efficacy (Chippaux and Goyffon 2008).

After a quick initial assessment and preliminary treatment at peripheral health facilities, patients with an exposure history of scorpion stings should be promptly referred to tertiary hospitals with intensive care unit (ICU) beds, oxygen, and antivenom facilities. Particularly for children, scorpion stings should be treated as an emergency with early therapy with prazosin followed by a timely referral (Prasad et al. 2011). The significant step for the management of scorpion stings is the alleviation of unbearable pain. The local anesthetic is the most effective analgesia, administered by digital block in stings on fingers and toes (Prasad et al. 2020). At the referral center, early detection of cardiopulmonary complications such as myocarditis left ventricular failure, and cardiogenic shock based on clinical manifestation, electrocardiographical abnormality, biomarkers, and echocardiography can help in the risk assessment of patients and facilitate decision-making in the management of scorpion sting cases (Warrell 2019).

The primary therapy for scorpion stings is an intravenous administration of equine ASA, raised against the venom of scorpion species prevalent in that region. For

DOI: 10.1201/9781003540816-5

example, in India, the commercial ASA is produced against the venom of the Indian red Scorpion (*Mesobuthus tamulus*). In Iran, ASA is made against the venoms of the six endemic Iranian scorpions (*Androctonus crassicauda, Buthotus saulcyi, Buthotus schach, Odontobuthus doriae, Mesobuthus eupeus,* and *Heterometrus lepturus*) (Latifi and Tabatabai 1979). However, the effectiveness, purity, and insufficient quantity of scorpion venom-specific antibodies of the commercial ASA to effectively neutralize the scorpion venom are critical problems for successful antivenom therapy against scorpion stings.

According to Deshpande, Pandey, and Tiwari (2008), ASA can be polyclonal or species-specific. For instance, Anscrop, the first ASA approved in the USA, treats individuals stung by *Centruroides* scorpions (Traynor 2011). In contrast, Scorpifav is a polyvalent ASA used to treat stings by *Androctonus* and *Leiurus* species, while Saimr is an ASA used to treat stings by *Parabuthus* species (Carmo et al. 2015). The dose of ASA to be used for treating a patient depends on the symptoms caused by the amount of venom inoculation and the ASA's ability to neutralize the venom (Chippaux and Goyffon 2008).

5.2   LIMITATIONS OF ANTIVENOM THERAPY

Because most commercial ASAs are of lower grade, they may cause unfavorable effects. Additionally, ASA treatment may have drawbacks like poor effectiveness against scorpion venom. The adverse responses elicited by antivenom can be classified into early and late adverse reactions.

5.3   ANTIVENOM-INDUCED EARLY ADVERSE REACTIONS

The most severe early adverse responses occur within 24 hours following the use of antivenoms (León et al. 2013). As mentioned below, the early adverse reactions are mediated by three mechanisms.

5.3.1   IgE-Mediated Anaphylactic Reactions

Early adverse reactions, known as anaphylactic reactions, are caused by IgE antibodies in commercial antivenoms, including ASA. These antibodies have been found attached to basophilic or mast cell Fc receptors (FcRs). Specific antigens may cause the antibodies that are connected to cells to cross-link. Encourage the degranulation and release of pro-inflammatory cytokines, histamine, leukotrienes, and other pharmacological mediators during the initial phase. Many reactions are brought on by these compounds, including increased vascular permeability, vasodilatation, contraction of the smooth muscles in the belly and lungs, generation of mucus, and localized inflammation (Cruce and Lewis 2004). Anaphylactic shock, characterized by edema in different tissues and a drop in blood pressure due to vasodilation, can be brought on by antigens found in heterologous antivenoms or throughout the body (Abbas and Litchman 2003; de Silva, Ryan, and de Silva 2016).

5.3.2 Non-IgE-Mediated Anaphylactic Reactions (Anaphylactoid Reactions)

Most early responses brought on by antivenoms are non-IgE-mediated anaphylactic ones. Low-molecular-weight active peptides known as anaphylotoxins (C3a, C4a, and C5a) are produced when the complement system is activated. The C3, C4, and C5 serum complement proteins are the source of anaphylotoxins, which are made of antigen–antibody complexes, immunoglobulin aggregates, and other substances cleave these proteins during complement fixation (Morais and Massaldi 2009).

Activating the conventional complement-mediated immunoglobulin aggregates is probably the crucial step in producing anaphylotoxins in the case of adverse reactions to antivenom (León et al. 2013; Abbas, Lichtman, and Pillai 2018). Additionally, heterophilic antibodies in antivenom against human neutrophils, erythrocytes, and other cell types may help produce anaphylatoxins (León et al. 2013). Chemotaxis and neutrophil activation are stimulated by the C5a, C3a, and C4a fragments, which cause basophils and mast cells to degranulate and release pharmacologically active mediators of acute hypersensitivity (Figure 5.1) (Abbas and Litchman 2003). Vascular smooth muscle is contracted, vascular permeability is raised, and neutrophils and monocytes migrate from the blood arteries as the overall results of these actions (Cruce and Lewis 2004). Patients without previously developed antivenom component hypersensitivity may display these symptoms (León et al. 2013).

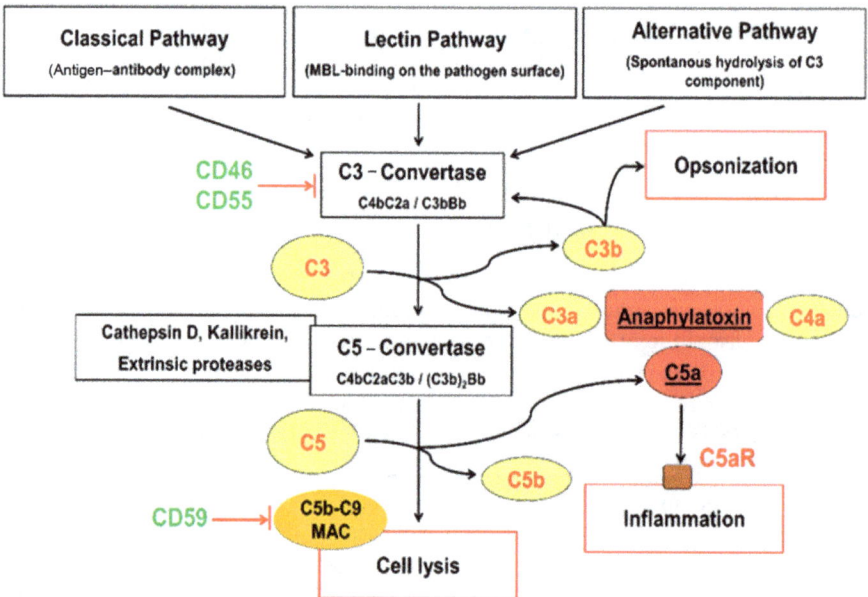

FIGURE 5.1 An overview of the complement activation pathways and biological effects mediated by complement products. (Figure and legend were adapted from Silawal et al. 2021; CC BY 4.0.)

Different antivenoms from different sources have been shown to activate the complement system's classical route and create anaphylatoxins (Squaiella-Baptistao et al. 2014). These findings imply that composition, contaminating proteins, and aggregation may all impact the anticomplementary action of antivenoms. Also hypothesized is independent mast cell activation brought on by non-complement activation (Figure 5.1) (Stone et al. 2013).

5.3.3 PYROGENIC REACTIONS

The primary source of pyrogenic reactions elicited by antivenoms is bacterial endotoxin contamination. Fortunately, endotoxin contamination is minimized by most production laboratories by implementing or starting to implement strict quality maintenance necessities for their facilities, raw materials, processing systems, and equipment. As a result, there has been a significant decline in this kind of adverse reaction in recent years.

Lipopolysaccharides (LPS) constitute bacterial endotoxins, a crucial component of Gram-negative bacteria's outer cell membranes (Erridge, Bennett-Guerrero, and Poxton 2002). Toll-like receptor 4 (TLR4) and/or LPS-binding protein (LPB) receptors are found on monocytes and other immune system cell components, which are responsible for producing TNF-, IL-6, interleukin 1 (IL-1), and other cytokines. These interactions are considered the molecular basis of toxicity (Morais and Massaldi 2009).

Endotoxin contamination may be observed in pharmaceutical products at low percentages, while contamination at higher levels is associated with bacterial infection or digestive system damage. Patients are more likely to experience minor responses (mainly fever) when small quantities of endotoxins are present in antivenom (Otero-Patiño et al. 1998). Last, preclinical evaluation of antivenoms in light of the 3Rs (replacement, reduction, and refinement) is required to prevent adverse effects in patients, particularly contamination by microbes (Gutiérrez et al. 2017).

5.3.4 LATE ADVERSE RESPONSES, TYPE III HYPERSENSITIVITY, OR SERUM SICKNESS

"Serum sickness" refers to late adverse responses 5–20 days after receiving antivenom (León et al. 2013). Pirquet and Schick were the first to document this unfavorable reaction in 1905 (Silverstein 2000). They discovered that several days after receiving massive doses of antitoxins, some patients experienced fever, rashes, and renal impairment with proteinuria and lymphadenopathy. Additionally, these scientists discovered that symptoms developed more quickly following a second exposure to the foreign serum than the initial dosage (Williams, Habib, and Warrell 2018).

Antigen–antibody complexes act as the mediating factor in type III hypersensitivity. The patient's immune system responds to the administration of antivenom by producing antibodies that adhere to the antivenom, which causes the development of immune complexes (Morais and Massaldi 2009; Morais et al. 2012). These complexes cause leukocyte infiltration and complement activation, resulting in the so-called "serum sickness" condition. The traditional reaction occurs 7–15 days after the triggering injection. In contrast, the fast version of serum sickness, which can occur in individuals already sensitized, may appear a few days after the injection. It has been difficult to ascertain how frequently this response happens because the symptoms are usually mild and appear after the patient has been released from the hospital without a medical record being made (Morais 2018).

The primary reason for the side effects caused by commercial antivenom is their impurity, having undergone only one or two purification steps. The contaminants cause (i) complement reactions, (ii) serum sickness, and (iii) anaphylactic shock. Specific alternative systems for treating scorpion sting envenomation have been introduced to avoid the interactions between ASA and immunoglobulin Fc regions and human complement.

5.4   HOW TO PREVENT ASA-INDUCED ADVERSE REACTIONS IN PATIENTS?

The following are some treatments clinicians frequently use to treat commercial ASA-induced adverse reactions in patients.

5.4.1   ANTIHISTAMINES

The amino acid L-histidine is decarboxylated to produce histamine, a key mediator of acute allergic and inflammatory responses, but has a minor function in anaphylaxis. Mast cells and basophils produce the majority of tissue histamine in their granules (Katzung and Julius 2001). Histamine's main effects include edema, decreases in systolic and diastolic blood pressure, increased heart rate, stimulation of sensory nerve endings, especially those mediating pain and itching, and bronchoconstriction in asthmatic patients. It also has a direct vasodilator effect on arterioles and precapillary sphincters (Katzung and Julius 2001).

Additionally, due to the active chemotactic attraction that histamine has for immune cells like neutrophils, eosinophils, basophils, monocytes, and lymphocytes, leakage of plasma containing acute inflammation mediators like complement proteins and C-reactive protein as well as antibodies occurs (Katzung and Julius 2001). Histamine interacts with particular cellular receptors on the membrane surface to produce biological effects. H1–H4 refers to the four distinct histamine receptors (Bullock and Manias 2013).

Immune reactions include the H1 receptor; antihistaminic H1 compounds treat or prevent allergic response symptoms. The most effective medications for treating urticaria are H1 antagonists, which are particularly beneficial when given before exposure. Histamine is the primary mediator of urticaria. H1 antagonists, however, are ineffective in illnesses involving several mediators, such as bronchial asthma, another pathology. There are two generations of H1 antagonists: first and second. Reversible competitive binding to the H1 receptor reduces or blocks histamine's effects (Bullock and Manias 2013).

The most often utilized premedications against antivenom-induced adverse reactions are promethazine and chlorpheniramine, both first-generation agents (de Silva et al. 2011). The actions of other mediators, such as prostaglandins and leukotrienes, are unaffected by antihistamine premedication, which can theoretically block or diminish the adverse effects of histamine but not those of other mediators.

5.4.2 GLUCOCORTICOIDS

Glucocorticoids have far-reaching impacts because they affect several cell types and metabolic processes throughout the body (Chrousos 2012). Such an influence may have significant consequences related to the side effects of this type of drug. Most glucocorticoid effects are mediated through broadly distributed glucocorticoid receptors. They govern the transcription of particular target genes, which impacts the regulation of growth factors, pro-inflammatory cytokines, and other factors (Chrousos 2012).

Regarding their immunological effects, glucocorticoids significantly lessen the outward signs of inflammation. The suppression of phospholipase A and cyclooxygenase activity as well as the prevention of the production of inflammatory and immunological mediators are some of the mechanisms behind this process. By causing an increase in the production of the intracellular mediator annexin-1, glucocorticoids inhibit the enzyme phospholipase A (Bullock and Manias 2013). Other immunosuppressive effects include a reduction of the lymph nodes and spleen, inhibition of helper T cells, a reduction in the production of antibodies and cytokines, a decline in neutrophil accompanied by macrophage phagocytic activity, and stabilization of membranes of mast cells, which lessens basophil and mast cell histamine release. Additionally, glucocorticoids alter the usual distribution of immune cells; more neutrophils circulate, whereas numbers of monocytes, eosinophils, basophils, T and B cells, and lymphocytes decline (Chrousos 2012).

Cortisone and hydrocortisone, which are natural glucocorticoids, also exhibit mineralocorticoid actions. Glucocorticoids are crucial therapeutic agents in managing various inflammatory, immunologic, and hematologic illnesses (Chrousos 2012). Short-, intermediate-, and long-acting versions of glucocorticoids are categorized by how long they remain active (Bullock and Manias 2013). In treating antivenom, the natural short-acting glucocorticoid hydrocortisone is frequently used as a premedication (de Silva, Ryan, and de Silva 2016). Glucocorticoids should affect phospholipase A and cyclooxygenase inhibition and mast cell membrane stabilization, essential for preventing anaphylactic responses. The decrease in antibody production should also mitigate late adverse reactions. Unfortunately, glucocorticoids' additional immunosuppressive effects take longer to manifest, thus rendering them useless as a preventative medicine for early negative responses (de Silva, Ryan, and de Silva 2016).

5.4.3 CATECHOLAMINES

Adrenaline (epinephrine) is the most often used catecholamine drug to treat or prevent early adverse reactions to antivenoms (Morais 2018; Mukherjee 2021). In contrast to glucocorticoids and antihistamines, harmful reaction mechanisms are not affected by adrenaline. Intense activities exhibited by adrenaline immediately counteract the effects of mast cell and basophil activation. It is a potent vasoconstrictor and cardiac stimulant because it is an agonist at the α and β adrenoceptors. Vascular beds contain many α1 receptors, and their activation causes venous and arterial vasoconstriction. The heart's receptors are stimulated, which results in increasing cardiac output. Notably, bronchodilation (expansion of bronchial airways) results from activating β_2 receptors in the bronchial smooth muscle (Biaggioni and Robertson 2012).

Adrenaline also affects many tissues and organs, including the eyes, pancreatic islets, salivary glands, apocrine sweat glands, genitourinary organs, fat cells, liver, and other endocrine glands (Biaggioni and Robertson 2012). Many clinicians prefer adrenaline exclusively to treat acute adverse reactions and not pretreatment due to its potent and widespread activities(de Silva et al. 2011; de Silva, Ryan, and de Silva 2016). The symptoms of anaphylactic shocks, which include bronchospasms, mucous membrane congestion, angioedema, and severe hypotension, often respond quickly to parenteral adrenaline treatment (Biaggioni and Robertson 2012).

5.4.4   USE OF PRAZOSIN: AN α-ADRENERGIC RECEPTOR BLOCKER

A postsynaptic α_1 blocker called prazosin (adrenoreceptor antagonist) combats the effects of excessive catecholamine and stops the emergence of severe systemic characteristics. According to Bawaskar and Bawaskar (2007), it has been discovered to be an effective medication for treating scorpion sting envenomation, and it has decreased the fatality rate to 1% from 30% in the pre-prazosin period (Bawaskar and Bawaskar 2007).

Particularly in India, prazosin is recommended for treating scorpion stings (Bawaskar and Bawaskar 1996). This medication is simple to use and has few significant side effects. Prazosin is more efficient than nifedipine, which prevents calcium ion inflow into the arterioles and inhibits the contraction of smooth muscle cells (Bawaskar and Bawaskar 1994). Hydralazine prevents the release of calcium ions from the vascular wall's smooth muscle. Despite being successful, hydralazine has several drawbacks, such as sympathetic stimulation that raises heart rate and increases the risk of myocardial infarction, as well as a rise in plasma renin that causes urinary retention and necessitates using a diuretic and α_2-adrenergic blocker. Additionally, parenterally administered hydralazine causes a protracted hypotensive response that is challenging to regulate (Ismail 1995).

Another approach is the medication captopril, which prevents angiotensin from being converted (Krishnan, Sonawane, and Karnad 2007). However, Captopril prevents the breakdown of bradykinin, which is crucial for the emergence of acute pulmonary edema (Ismail 1995). The agonist of α_2 adrenergic receptors, clonidine, reduces sympathetic activity. Consequently, both peripheral blood pressure and heart rate are reduced. Although clonidine may be helpful, it has never been evaluated in treating scorpion envenoming (Bawaskar and Bawaskar 1996).

5.5   IN VITRO AND IN VIVO QUALITY ASSESSMENT OF ASA: AN EVALUATION TO DETERMINE THE EFFICACY OF COMMERCIAL ASA

The success of managing scorpion stings or venomous bites in patients is mainly contingent upon the safety, effectiveness, and consistency of the formulation of ASA

(Das et al. 2022). Antivenoms are the only medication for neutralizing toxins in animals' venom, such as scorpions and snakes, through antigen–antibody binding. Several analyses are conducted throughout its production to ensure the quality and effectiveness of the antivenom administered to the patient. One is the potency assay, which evaluates antivenom's ability to neutralize the venom's toxic effects in mice (WHO 2019). The substitution of in vivo for in vitro assays such as ELISA (Enzyme-Linked Immunosorbent Assay) has been presented by other authors, bringing several advantages, such as the reduction in the use of animals, in costs, and the duration of the assays (Silva et al. 2023).

Through in vitro laboratory testing, the safety, effectiveness, and homogeneity of commercial ASAs made in India were evaluated (Das et al. 2022). The study showed that the ASA was prepared according to the guidelines of the World Health Organization. The ASAs were devoid of aggregate content and virus particles. Biophysical analyses demonstrated the poor binding of venom with commercial ASAs. However, commercial ASAs contain less proportion of scorpion venom toxin-specific antibodies (Das et al. 2022).

The neutralization capacity of monovalent *A. crassicauda* antivenom (RSIIA anti-Ac) was tested in Swiss mice (both sexes) against existing lethal scorpions in the Turkish scorpio fauna such as *M. eupeus, Mesobuthus gibbosus,* and *A. crassicauda.* The neutralization capacity of one mL ASA was found to be against 23 LD_{50} of *M. eupeus*, 32 LD_{50} of *M. gibbosus*, and 42 LD_{50} of *A. crassicauda* venoms (Ozkan and Yağmur 2017). Another study also evaluated the avidity of sera from hyper-immunized horses with crude *Tityus serrulatus* venom, a scorpion species associated with the most severe accidents in Brazil, and its potential for application as a potency test replacing the in vivo assay (Silva et al. 2023).

When given shortly after a sting, ASA is a particular antidote that can neutralize circulating venom toxins. Skilled medical professionals in Brazil, Mexico, Tunisia, Venezuela, and Iran have found success with ASA (Freire-Maia, Campos, and Amaral 1994; Ismail 1995; Ghalim et al. 2000; Dehesa-Dávila and Possani 1994; de Dàvila et al. 2002). Nevertheless, serotherapy for scorpion stings has been controversial (Abroug et al. 1999; Sofer, Shahak, and Gueron 1994; Bawaskar and Bawaskar 1991; Boyer et al. 2009). It remains uncertain if the ASA can reverse the pathophysiological consequences of scorpion venom on the heart.

Numerous investigations have demonstrated that commercial ASA does not reduce hemodynamic alterations, cardiogenic pulmonary edema, or prevent death (Sofer, Shahak, and Gueron 1994; Cupo and Hering 2002) and that both ASA-treated and non-treated individuals had the same outcomes (Abroug et al. 1999; Bawaskar and Bawaskar 1992). De Rezende and colleagues demonstrated that patients with pulmonary edema recovered only 48 hours following serotherapy. But an hour following ASA treatment, there was no longer any evidence of venom antigen in the plasma of people stung by a scorpion, and agitation and pain subsided in a few hours (De Rezende et al. 1995).

Since 2002, monospecific ASA consisting of $F(ab)_2$ has been prepared by immunizing horses against scorpion venom. It has been available for clinical use from some commercial ASA manufacturing companies in India. Prazosin, an α-adrenergic inhibitor, is frequently used to treat *M. tamulus* stings (Bawaskar and Bawaskar 1986;

Al-Asmari et al. 2008; Bawaskar and Bawaskar 2000; Gupta 2006; Bosnak et al. 2009; Peker et al. 2010; Koseoglu and Koseoglu 2006; Warrell 1997).

Severe *M. tamulus* stings were treated using a prospective randomized trial comparing the use of scorpion antivenom in combination with prazosin alone. The amount of venom, the size of the scorpion, the season, the victim's age, and the interval between the sting and hospital admission all affect how severe a scorpion sting is. Since its introduction for clinical usage in India in 2002, ASA has been scarce and difficult to get in pharmacies; in contrast, prazosin is widely accessible (Bawaskar and Bawaskar 2000).

Scorpion stings typically occur at night, at the victim's house, in rural areas, small towns or suburbs, and in the city's heart. If the patient was referred from a health facility lacking adequate resources for treatment, or if they initially preferred traditional therapy, their arrival at the health center may be delayed by up to two hours following the sting (Chippaux and Goyffon 2008; Goyffon, Vachon, and Broglio 1982; Soulaymani-Bencheikh et al. 2002; Celis et al. 2007). First aid is typically provided by medical personnel who have received insufficient training.

5.5.1 TREATMENT FOR SYMPTOMS AND ADJUVANTS

Even though they are not the most essential medications for envenoming, using analgesics is crucial because pain is frequent and severe. Salicylates and other anti-inflammatory drugs can be helpful, but use caution as doing so may cause Reye's or Lyell's syndrome, which is more likely in children receiving salicylate treatment for viral infections (Arrowsmith et al. 1987). Reye's syndrome has a modest incidence— 0.8 cases per million children—but it frequently results in death because its symptoms are similar to those of progressive envenoming. As 0.5–1% of scorpion stings in children can be deadly, it is essential to weigh the risks and benefits of various therapies while also considering that 15–20% of scorpion stings might proceed severely. According to some clinicians, 1% of lignocaine should be used for local anesthetic purposes (Ismail 1994). However, although morphine and its analogs, such as codeine and tramadol, are very effective (Nascimento Jr et al. 2005), these drugs should be avoided because they worsen respiratory depression in patients and inhibit noradrenaline reuptake, which can amplify their effects.

Numerous medications targeting the heart have been proposed to treat pulmonary edema, heart failure, arrhythmia, and hypertension linked to scorpion envenomation. However, keep in mind that the primary source of these symptoms is the peripheral vasoconstriction brought on by catecholamine activity, which increases vascular resistance. Prazosin is remarkably advised in India for treating scorpion envenomation (Bawaskar and Bawaskar 1994). This medication has few major contraindications and is simple to use. Prazosin is more efficient than nifedipine, which prevents calcium ion inflow into the arterioles and inhibits the contraction of smooth muscle cells (Bawaskar and Bawaskar 1996).

Hydralazine prevents the release of calcium ions from the vascular wall's smooth muscle. Despite being successful, hydralazine has several drawbacks, such as sympathetic stimulation that raises heart rate and increases the risk of myocardial infarction, as well as a rise in plasma renin that causes urinary retention and necessitates using a

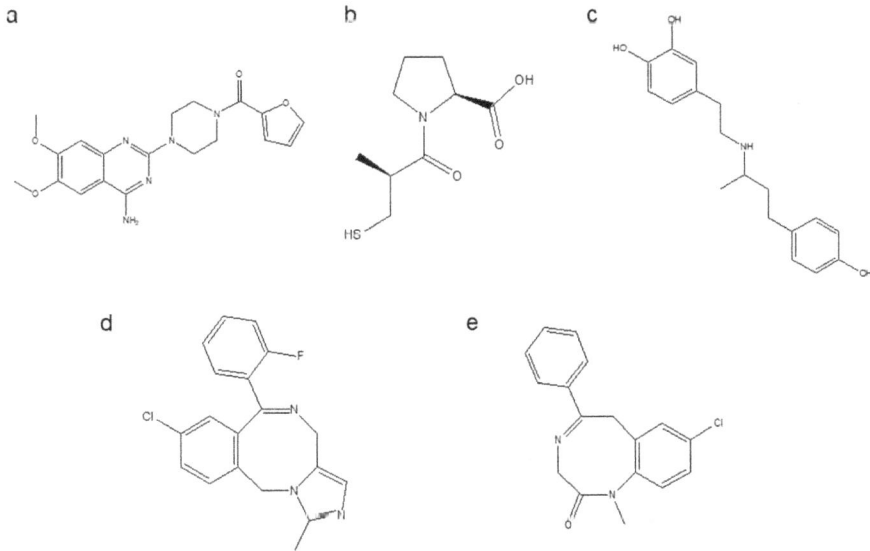

FIGURE 5.2 Chemical structure of some common drugs used to treat scorpion stings. a. Prazosin, b. Captopril, c. Dobutamine, d. Midazolam, e. Diazepam. Figures were drawn using ChemDraw 8.0 software.

diuretic and α_2-adrenergic blocker (Nath et al. 2022). Additionally, parenterally given hydralazine causes a protracted hypotensive response that is challenging to regulate (Ismail 1995). Figure 5.2 (a–e) shows different drugs with chemical structures for treating scorpion stings.

Another option is the medication captopril, which prevents the conversion of angiotensin (Krishnan, Sonawane, and Karnad 2007). On the other hand, Captopril prevents bradykinin from degrading, which is a crucial factor in developing acute pulmonary edema (Ismail 1995). An α_2-adrenergic receptor agonist, clonidine reduces sympathetic activity. Heart rate and peripheral blood pressure both decrease as a result. To the author's knowledge, clonidine has not been studied yet to treat scorpion envenoming, although it may be a helpful medication.

Several physicians use dobutamine alone or with diuretics or antiarrhythmics when heart failure is proven. Prazosin has been shown (Gupta et al. 2007) to have a slight advantage over dobutamine regarding recovery time and ease of administration. Heart failure is a late consequence of scorpion envenoming; however, this could be because of the specific conditions or type of scorpion involved. Drugs that activate gamma-butyric acid (GABA) receptors, which reduce postsynaptic neurons' excitability, are frequently used to treat neuromuscular diseases. Because of their quick distribution in the body and brief half-life, benzodiazepines have an advantage over other anticonvulsants like barbiturates (phenobarbitone). The respiration is also depressed by barbiturates.

The most popular drug, especially in North America, is midazolam (Boyer et al. 2009; Gibly et al. 1999). Diazepam is widely used elsewhere, notably in Africa and the Middle East (Ismail 1994). Additionally, benzodiazepines help manage hypertension and may be the first therapeutic option for scorpion envenoming. Antiparasympathetic medications like atropine are often not advised to treat scorpion envenoming. These prevent sweating, which is crucial for controlling body temperature, especially in children, and they increase the adrenergic effects of scorpion venom, which raises the risk of ischemia problems and hypertension (Freire-Maia 1989; Ismail 1994; Bawaskar and Bawaskar 1992; Seifert 2001). But in cases of entire atrioventricular block or severe bradycardia, which are usually found, these drugs might be helpful.

Anti-inflammatory medications to treat scorpion envenoming have never been the subject of official clinical investigations. Even with the possibility of adverse effects, there is no question about the therapeutic usefulness of these drugs. As we have seen, experimental studies have demonstrated the involvement of several biological indicators of inflammation in scorpion envenoming, and clinical trials have supported these findings. There are, therefore, new issues for clinical research. Salicylates are effective analgesics with anti-inflammatory action, making it easy to control pain (Amann and Peskar 2002). Choosing the best medication is difficult because of the wide range of indications and dosages that depend on venom composition, which varies between species and even individuals; the amount of venom injected; complications associated with the progression of envenoming, like delayed consultation; drug side effects; and modes of drug administration, especially in outlying health centers where physicians are not always present (Nath and Mukherjee 2022).

5.6   PASSIVE IMMUNOTHERAPY

Passive immunotherapy, developed in 1894, is the only etiological treatment for scorpion envenomation and snakebite (Chippaux and Goyffon 1998). The basis of immunotherapy is the injection of antibodies generated from an animal that has previously developed a hyperimmune response to the venom of the same or a closely related species. Enhancing antivenomous sera involves isolating the antibodies from other plasma components first, then IgG is enzymatically digested, and finally, the finished product is purified. As a result, effectiveness and safety are greatly improved (Mukherjee 2021). Most antivenoms now produced are purified IgG F(ab')$_2$ fragments, which reduces possible adverse reactions. However, poorly prepared antivenoms might result in dangerous side effects, including shock or anaphylaxis, which call for the quick injection of adrenaline (Morais 2018).

To enhance the venom's diffusion and simplify its complexation, antivenom should be injected intravenously as soon as practical. IgG antibodies and F(ab')$_2$ fragments attached to the blood compartment's venom do not diffuse. After removing the antigen–antibody combination that the venom and F(ab')$_2$ created, the venom is extracted from the tissue compartment and added to the blood, which is then complexed with F(ab')$_2$. Even if antibodies do not instantly neutralize the venom, removing the toxin produces a rapid regression of symptoms. Both experimental and clinical research have supported this (Boyer et al. 2009; Ghalim et al. 2000; Hammoudi-Triki et al. 2004; Sevcik et al. 2004). The symptoms determine the dosage, the interval between the sting and the initiation of treatment, the clinical development, the titer of the antivenom, and the medical environment, including the ability to provide adjuvant therapy. One of the primary factors that has to be documented is probably dosage.

ASAs are especially contentious in Africa and India for several reasons, including their efficacy and application in treating scorpion envenomation (Soulaymani-Bencheikh et al. 2002; Soulaymani Bencheikh et al. 2007; Bawaskar and Bawaskar 2007; Abroug et al. 1999). First, there is a worry about adverse reactions from using whole IgG that has not been appropriately refined. The high incidence of side effects (occurring in at least 50% of patients treated with anti-venom), some of which are life-threatening, offsets the low case fatality rate associated with scorpion stings (1–2%) at the time. This problem is no longer the case when introducing $F(ab')_2$, whose side effects are rare (5%) and usually mild (Chippaux and Goyffon 1998; Chippaux and Goyffon 2008). Second, there has been concern about incorrect ASA administration, probably by an insufficient dose. The majority of treatments, including those used in clinical trials, employed antivenoms with a neutralizing titer of around 10 ED_{50} (i.e., 1 mL of antivenom neutralized an amount of venom equivalent to 10 LD_{50}), whereas the most recent antivenoms exhibit a neutralizing titer of more than 50 ED_{50}, i.e., they are 3–5 times more effective. Finally, relatively few and frequently biased clinical trials have been conducted to verify antivenoms' effectiveness (Chippaux, Stock, and Massougbodji 2010).

However, a randomized, placebo-controlled, anonymous clinical trial conducted by Boyer and his co-workers (Boyer et al. 2009) revealed that antivenom significantly reduced the amount of benzodiazepine required in the group treated with antivenom. All patients treated with antivenom attained that cure in less than 4 hours (versus 15% in the control group). In less than an hour, the antivenom eliminated the venom from the plasma compartment (compared to an average of over 4 hours for the placebo group).

5.7 IMPROVED TREATMENT OF SCORPION ENVENOMATION

The substitution of in vivo for in vitro assays has been presented by some authors, bringing several advantages, such as the reduction in the use of animals, in costs, and the duration of the assays (Silva et al. 2023). The limitations of traditional treatments utilizing α_1-adrenoreceptor antagonists and commercial ASA have led the writers to look for a suitable formulation to enhance care for Indian red scorpion *M. tamulus* stings. When evaluated in *Caenorhabditis elegans* and Wistar strain albino rats in vivo models, novel therapeutic medication formulations containing low dosages of commercial ASA, AAA, and ascorbic acid have significantly improved in neutralizing the in vivo toxic effects of *M. tamulus* venom (MTV) (Das et al. 2023).

The Butantan Institute manufactures the following two medicinal scorpion venoms in Brazil: (i) an antiarachnidic antivenom is produced by immunizing horses with a combination of venoms from *T. serrulatus* (57%), *Phoneutria nigriventer* (21.5%), and *Loxosceles gaucho* (21.5%); and (ii) immunizing horses produce ASAs with a combination of venoms from *T. serrulatus* (50%) and *T. bahiensis* (50%). The therapeutic ASAs may be ineffective in recognizing the critical components of various venom species found nationwide due to the probable variability of composition and toxicity of *Tityus* spp. venom (Venancio et al. 2013). This study aimed to evaluate the antigenic cross-reactivity of Brazilian ASAs to assess their capacity to neutralize the enzymatic activities of venoms derived from *T. serrulatus*, *T. bahiensis*, and *T. stigmurus*, as well as to characterize the enzymatic properties of these venoms (Cupo 2015).

5.8 TRADITIONAL MEDICINES IN THE TREATMENT OF SCORPION STINGS: A BIRD'S EYE VIEW

Herbs and medicinal plants are the mainstay of popular ethnomedical treatments for scorpion stings. Since ancient times, herbal remedies have been vital; today, many people, particularly in underdeveloped nations, use ethnomedicinal plant treatments to treat diseases. Traditional medical practices are in greater demand than ever as they are cheap, plentiful, safe, and effective in controlling infections and enhancing health (Bahekar, Kale, and Nagpure 2012; Satpute 2019).

Therefore, long-established practices, although not supported by the health authorities of several countries, are relied upon by 80% of the population. Ayurveda, a conventional form of Indian medicine, and other complementary therapies are used by 65% of the country's rural residents to cure a variety of illnesses, including deadly venomous snake and scorpion stings (Bahekar, Kale, and Nagpure 2012; Binorkar 2012; Mukherjee and Chattopadhyay 2022; Das et al. 2023; Puzari, Fernandes, and Mukherjee 2022; Nath and Mukherjee 2022). Most healing herbs are designed to treat basic ailments like pain alleviation and inflammation control. Some are given orally to neutralize the venom, while others are used as an antidote and antivenom. Some are recommended as juice, powder, paste, poultice, or latex. An extensive list of several ethnomedicinal plants that have been used to cure scorpion stings throughout history is provided in Table 5.1.

5.9 THE PROS AND CONS OF USING MEDICINAL PLANTS FOR TREATING SCORPION VENOM

People in earlier times had limited alternatives but relied on traditional remedies made from locally accessible ethnomedicinal plants, animals, magical drugs, and even goods made from scorpion venom (González and Vallejo 2013). The discrepancy between traditional practices and current understanding precludes approval for clinical use, even though these conventional practices are generally regarded as relatively safe and effective, but with limited scientific support. There are remarkably few studies that examine the pharmacological characteristics of medicinal plants and their ability to neutralize scorpion stings in vivo and in vitro. Few studies have examined unpurified bioactive components' phytochemical characteristics and how well they protect against scorpion stings. There is also a dearth of information on the pharmacological effects of therapeutic plants.

Due to three primary factors, modern physicians do not accept using herbal medicines to treat scorpion stings. First, they are less efficient in reducing severe symptoms and are only beneficial in reducing early signs of scorpionism. Second,

TABLE 5.1
A list of herbs that have been used historically to treat scorpion stings

Sl. No	Scientific Name	Common Name	Family	Country	Traditional Route of Administration	Reference(s)
Traditional use in the Indian subcontinent						
1	*Abutilon indicum* (L.) Sweet	Monkey Bush	Malvaceae	Andhra Pradesh, India	Leaf paste is applied over the place of the scorpion sting.	(Krishna, Saidulu, and Hindumathi 2014)
2	*Achyranthes aspera* L.	Aghada, Kutra, *Vuttareni*	Amaranthaceae	Maharashtra, India	5 g roots are boiled in 100 mL water, and 50 mL extract is prepared; the whole extract is then given once as an antidote.	(Satpute 2019)
3	*Achyranthes aspera* L.	*Vuttareni*	Amaranthaceae	Andhra Pradesh, India	Fresh leaves of the plant are ground and applied on the site of the scorpion sting.	(Basha and Murthy 2017; Butt et al. 2015)
4	*Achyranthes aspera* L.	Prickly chaff flower	Amaranthaceae	Pakistan	The paste of leaves is applied to the scorpion bite.	(Nasim et al. 2013)
5	*Achyranthes aspera* L.	Geshay	Amaranthaceae		Fresh leaves of the plant are ground and applied on the bite site for one hour.	(Butt et al. 2015)
6	*Acorus calamus* L.	SkhaWaja	Acoraceae	Northern Pakistan, Andhra Pradesh, India	The dried rhizome is taken to make powder and applied to the infected site.	(Basha and Murthy 2017)
7	*Adiantum capillus veneris* L.	Maidenhair fern	Adiantaceae	Pakistan	Leaves are used for the scorpion stings.	(Nasim et al. 2013)

(continued)

TABLE 5.1 (Continued)
A list of herbs that have been used historically to treat scorpion stings

Sl. No	Scientific Name	Common Name	Family	Country	Traditional Route of Administration	Reference(s)
8	*Adiantum capillus-veneris* L.	Perishoan	Pteridaceae	Northern Pakistan	Fresh fronds are taken and crushed to make a paste. The paste is then applied on the bite site for2–3 times a day.	(Butt et al. 2015)
9	*Adiantum venustum* D. Don	Baboza	Pteridaceae	Northern Pakistan	Fresh fronds are taken and crushed to make a paste. The paste is then applied on the bite site 2–3 times daily.	(Butt et al. 2015)
10	*Agave americana* L.	Century plant	Agavaceae	Spain	A poultice is applied on the site of the sting.	(Gonzalez and Vallejo 2013)
11	*Ageratum houstonianum* Mill.	Neelibooti	Asteraceae	Northern Pakistan	A few drops of water are used to grind 200 g of leaves and flowering parts from this mixture juice is extracted using a thin cloth. 5–7 drops of this juice are poured on the wound.	(Butt et al. 2015)
12	*Allium cepa* L.	Piaz	Amaryllidaceae	Northern Pakistan	Leaves or mostly fresh bulbs are ground without any solvent, and dense juice is obtained, which is applied on the sting site.	(Butt et al. 2015)
13	*Allium cepa* L.	Onion	Alliaceae	Saudi Arabia	Bulbs are carefully washed and then fried. The hot bulbs are applied to the sting site. The dose is once only to release the contaminated blood.	(Fakhry, Migahid, and Anazi 2017)

No.	Scientific name	Common name	Family	Country/Region	Description	Reference
14	*Allium cepa* L.	Onion	Liliaceae	Pakistan	Bulb and leaves are taken.	(Nasim et al. 2013)
15	*Allium sativum* L.	Garlic	Liliaceae	Spain	Crushed garlic is applied as a poultice on the stigma of the sting	(Gonzalez and Vallejo 2013)
16	*Amaranthus caudatus* L.	Chalway	Amaranthaceae	Northern Pakistan	Poultices are made by grinding fresh leaves and applied on the infected site.	(Butt et al. 2015)
17	*Amaranthus virdis* L.	Slender amaranth	Amaranthaceae	Pakistan	The paste of the root is applied to the scorpion sting	(Nasim et al. 2013)
18	*Amaranthus viridis* L.	Chalveray	Amaranthaceae	Northern Pakistan	About 300 g of plant material is used to make a paste. This paste is applied on the bite site 2 or 3 times daily.	(Butt et al. 2015)
19	*Anacyclus pyrethrum* (L.) Lag. and *Opoponax galbanum*	Akarkara	Asterales	Iran	Combination of garlic, warm wine bottle grass, angeda resin, and Opaponaxgalbanium is given orally.	(Dehghani and Arani 2015)
20	*Andrographis paniculata* (Burm.f.) Wall	Kalmegh	Acanthaceae	Maharashtra, India	Aerial parts – ethanolic extract for venom neutralization	(Dnyaneshwar and Sonali 2016)
21	*Andrographis paniculata* (Burm.f.) Wall	Nelavemu	Acanthaceae	Andhra Pradesh, India	Pills are made from the leaves of the plant that are used as an antidote	(Basha and Murthy 2017)
22	*Annona squamosa* L.	Hitaphar	Annonaceae	India	Root paste is taken for external applications. Also, root bark decoction is taken orally.	(Bahekar, Kale, and Nagpure 2012; Basha and Murthy 2017)
23	*Argemone mexicana* L.	Perammathandu, Bilayat	Papaveraceae	India, Pakistan	Latex and yellow juice of the plant are taken orally. Root paste is also applied for local applications.	(Bahekar, Kale, and Nagpure 2012; Nasim et al. 2013)

(continued)

TABLE 5.1 (Continued)
A list of herbs that have been used historically to treat scorpion stings

Sl. No	Scientific Name	Common Name	Family	Country	Traditional Route of Administration	Reference(s)
24	*Arisaema flavum* Schott	Marjarai	Araceae	Northern Pakistan	Juice of the fresh rhizome is extracted and applied to snake bites and scorpion stings.	(Butt et al. 2015)
25	*Aristolochia indica* L.	Eswaramooligai	Aristolochiaceae	India	The leaf juice is taken orally to treat scorpion stings. Roots paste is applied externally on the scorpion sting part.	(Satpute 2019; Bahekar, Kale, and Nagpure 2012; Dnyaneshwar and Sonali 2016)
26	*Artemisia absinthium* L.	Wormwood	Asteraceae	Spain	The juice of this plant is applied to the sting area.	(Gonzalez and Vallejo 2013)
27	*Artemisia biennis* Hook.f.	Tarkha	Asteraceae	Northern Pakistan	Fresh flowers are taken to make a paste. The fresh paste is used three times a day.	(Butt et al. 2015)
28	*Artemisia brevifolia* Wall.	Chau	Asteraceae	Northern Pakistan	A paste of leaves or flowers is rubbed on the infected site. Replace this poultice every three hours.	(Butt et al. 2015)
29	*Artemisia scoparia* Maxim.	Virgate wormwood	Asteraceae	Northern Pakistan	The whole plant is used to prepare a paste.	(Butt et al. 2015)
30	*Asphodelus tenuifolius* Cav.	Onionweed	Liliaceae	Pakistan	The paste of leaves	(Nasim et al. 2013)
31	*Astragalus spinosus* (Forssk.) Muschl.	Milkvetch	Fabaceae	Saudi Arabia	Fresh or dried leaves are crushed and boiled in milk or water. One glassful of the leaf decoction is taken orally once	(Fakhry, Migahid, and Anazi 2017)

#	Scientific name	Common name	Family	Location	Use	Reference
32	Azadirachta indica A. Juss.	Neem tree	Meliaceae	Pakistan	Bark, leaves, twigs, and seeds are used for the scorpion sting.	(Nasim et al. 2013)
33	Azadirachta indica A. Juss.	Neem	Meliaceae	India	The whole plant except the root is used against the scorpion sting.	(Bahekar, Kale, and Nagpure 2012; Satpute 2019)
34	Bacopa monnieri (L.) Wettst.	Neerisambraniaku, Brahmi	Scrophulariaceae	Andhra Pradesh, India	Powder of the plant is given for nervous debility and as a brain tonic.	(Basha and Murthy 2017)
35	Barleria cristata L.	Phool Kanda	Acanthaceae	Northern Pakistan	A fine powder is obtained by grinding 400 g of plant material. Use this powder after each hour on the infected site.	(Butt et al. 2015)
36	Boerhavia coccinea Mill.	Itsit	Nyctaginaceae	Northern Pakistan	Extract juice from the fresh plant. Pour 2 to 3 drops of this juice into the infected area after two hours to reduce pain.	(Butt et al. 2015)
37	Boerhavia procumbens Banksex Roxb.	Jungliitsit	Nyctaginaceae	Northern Pakistan	10–20 mL juice is extracted from fresh leaves. After every 2 hours, a few drops are poured on the infected site.	
38	Bryonia dioica Jacq.	Red bryony	Cucurbitaceae	Spain	Its root is used as a painkiller.	(Gonzalez and Vallejo 2013)
39	Calotropis procera (Aiton) Dryand.	Safed Rui, Mhatari Rui	Asclepiadaceae	India, Pakistan, Saudi Arabia	Latex is used for local applications.	(Bahekar, Kale, and Nagpure 2012; Fakhry, Migahid, and Anazi 2017; Nasim et al. 2013)

(continued)

TABLE 5.1 (Continued)
A list of herbs that have been used historically to treat scorpion stings

Sl. No	Scientific Name	Common Name	Family	Country	Traditional Route of Administration	Reference(s)
40	*Calotropis procera* (Aiton) Dryand.	Akk	Asclepiadaceae	Northern Pakistan	Fresh latex from leaves is rubbed on the bite site thrice a day.	(Butt et al. 2015; Basha and Murthy 2017)
41	*Cassia occidentalis* (L.) Rose	Coffee senna	Caesalpinaceae	Pakistan	Roots are taken to treat scorpion stings.	(Nasim et al. 2013)
42	*Cissus quadrangularis* L.	Pirandai	Vitaceae	India	Seed and leaves juice—orally and externally.	(Bahekar, Kale, and Nagpure 2012)
43	*Citrullus colocynthis* (L.) Schrad.	Bitter apple, colocynth	Cucurbitaceae	Saudi Arabia	Half the fresh fruit is applied externally at the sting point for half an hour until its color turns black.	(Fakhry, Migahid, and Anazi 2017)
44	*Citrus aurantifolia* (Christm.) Swingle	Key Lime	Rutaceae	Saudi Arabia	Fruit juice is applied directly to the sting point.	(Fakhry, Migahid, and Anazi 2017)
45	*Citrus x aurantium* (L.)	Turng	Rutaceae	Northern Pakistan	The seeds are crushed, and the paste is made with water. Apply this ointment to the infected area.	(Butt et al. 2015)
46	*Citrus limon* (L.) Osbeck	Lemon	Rutaceae	Spain	Lemon juice is used as a topical antidote.	(Gonzalez and Vallejo 2013)
47	*Cleome gynandra* L.	Shona cabbage	Capperdiceae	Pakistan	Leaves, seeds, and roots are used for scorpion stings.	(Nasim et al. 2013)
48	*Commiphoramyrrha* Engl.	African myrrha	Burseraceae	Saudi Arabia	Gum powder is mixed with salt or dates and applied directly at the sting point.	(Fakhry, Migahid, and Anazi 2017)

49	*Convolvulus arvensis* L.	Bindweed	Convolvulaceae	Spain	The juice of this plant is applied to the string.	(Gonzalez and Vallejo 2013)
50	*Cynodondactylon* (L.) Pers.	*Durva, Harali*	Poaceae	Maharashtra, India	Whole plants are chewed in the mouth, and the juice is taken inside the stomach to relieve pain.	(Satpute 2019)
51	*Cyperus niveus* Retz.	Sedge	Cyperaceae	Pakistan	Roots are used for scorpion stings.	(Nasim et al. 2013)
52	*Cyperus rotundus* L.	Thunga	Cyperaceae	Andhra Pradesh, India	Dried tubers are pasted and applied topically on the bitten site of the scorpion.	(Krishna, Saidulu, and Hindumathi 2014)
53	*Daphne gnidium* L.	flax-leaved daphne	Thymelaeaceae	Spain	Apply a piece of bark.	(Gonzalez and Vallejo 2013)
54	*Datura stramonium* L.	Jimson town weed, mad apple, moonflower, thorn apple	Solanaceae	Saudi Arabia	Leaves are dried, crushed, heated, and mixed with flour, then applied externally to the sting point.	(Fakhry, Migahid, and Anazi 2017)
55	*Desmodium elegans* (Lour.) Benth.	Chamkat,	Fabaceae	Northern Pakistan	The fresh roots are powdered, and the poultice is applied to wounds caused by scorpion or snake bites.	(Butt et al. 2015)
56	*Desmodiumgangeticum* (L.) DC.	Salwan	Fabaceae	Northern Pakistan	A poultice of leaves is tied on wounds to treat snake and scorpion bites.	(Butt et al. 2015)
57	*Echinopsechinatus* Roxb.	Kandyari	Asteraceae	Northern Pakistan	Roots are boiled in water and taken orally to cure scorpion bites.	(Butt et al. 2015)
58	*Ecliptaprostrata*Lour.	Bhangra	Asteraceae	Northern Pakistan	Whole plant extract is given orally twice or thrice daily for seven days or until the patient fully recovers.	(Butt et al. 2015; Basha and Murthy 2017)

(continued)

TABLE 5.1 (Continued)
A list of herbs that have been used historically to treat scorpion stings

Sl. No	Scientific Name	Common Name	Family	Country	Traditional Route of Administration	Reference(s)
59	*Eryngium campestre* L.	Field eryngo	Apiaceae	Spain	Apply as a poultice on the sting. It very effectively alleviates the pain.	(Gonzalez and Vallejo 2013)
60	*Euphorbia caducifolia* Haines	Leafless milk hedge	Euphorbiaceae	Pakistan	The milky juice of leaves is used for the scorpion sting.	(Nasim et al. 2013)
61	*Euphorbia thymifolia* L.	Choti- dudhi	Euphorbiacea	Pakistan	Whole plant juice is used for scorpion bites.	(Nasim et al. 2013)
62	*Ferula narthex* Boiss.	Asafoetida	Umbelliferae	Pakistan	The gum resin is used.	(Nasim et al. 2013)
63	*Ficus carica* L.	Common fig	Moraceae	Spain	Apply latex to the area of the sting.	(Gonzalez and Vallejo 2013)
64	*Gymnemasylvestre* (Retz.) R.Br. ex Sm.	Podapathri	Asclepiadaceae	Andhra Pradesh, India	The root paste is applied to the scorpion sting.	(Krishna, Saidulu, and Hindumathi 2014)
65	*Heliotropiumbacciferum* Forssk.	Salt heliotrope	Boraginaceae	Saudi Arabia	The decoction of leaves or the whole shoot system is prepared with adequate boiling water. One glassful is taken orally once. The remaining boiled decoction is applied externally to the sting point.	(Fakhry, Migahid, and Anazi 2017)
66	*Hyoscyamus niger* L.	Henbane	Solanaceae	Spain	Oil is applied to the sting	(Gonzalez and Vallejo 2013)
67	*Jasminum officinale* L.	Chambeli	Oleaceae	Northern Pakistan	Fresh leaves are crushed to paste and applied on the bite site.	(Butt et al. 2015)

68	*Juniperus oxycedrus* L.	Brown Berried	Cupressaceae	Spain	A few drops of oil are applied to the affected area	(Gonzalez and Vallejo 2013)
69	*Justicia adhatoda* L.	Baikhar	Acanthaceae	Northern Pakistan	Half-cup of juice is taken orally twice a day. Leaves are also used and applied as poultice.	(Butt et al. 2015)
70	*Lagenaria siceraria* (Molina) Standl.	Bottle gourd	Cucurbitacae	Pakistan	Pulps of fruit are used for scorpion stings.	(Nasim et al. 2013)
71	*Linumusitatissimum* L.	Linseed, Flax	Linaceae	Saudi Arabia	Seed powder is mixed in water and applied directly to the sting point.	(Fakhry, Migahid, and Anazi 2017)
72	*Mangifera indica* L.	Aam	Anacardiaceae	Northern Pakistan, Andhra Pradesh, India	One teaspoon of the powder of dried leaves is used 2–3 times a day.	(Butt et al. 2015; Basha and Murthy 2017)
73	*Mangifera indica* L.	Amba,Mamarum	Anacardiaceae	India	Powder of flowers for local application	(Bahekar, Kale, and Nagpure 2012)
74	*Melia azedarach* L.	Hindustani shandai	Meliaceae	Northern Pakistan, Andhra Pradesh, India	A poultice of plant parts is made and applied to neutralize the toxic effect of scorpion bite.	(Butt et al. 2015; Basha and Murthy 2017)
75	*Mentha pulegium* L.	Penny royal	Lamiaceae	Iran	Syrup of the plant is taken to treat scorpion stings.	(Dehghani and Arani 2015)
76	*Nerium oleander* L.	Kanair,	Apocynaceae	Northern Pakistan	The paste is applied at the site of the sting.	(Butt et al. 2015)
77	*Ocimum basilicum* L.	Sweet Basil	Lamiaceae	Pakistan	Leaves are applied at the site of the sting.	(Nasim et al. 2013)

(continued)

TABLE 5.1 (Continued)
A list of herbs that have been used historically to treat scorpion stings

Sl. No	Scientific Name	Common Name	Family	Country	Traditional Route of Administration	Reference(s)
78	Ocimum tenuiflorum L.	Tulsi	Lamiaceae	India	Panchang and root are used for the scorpion sting.	(Gonzalez and Vallejo 2013)
79	Oenanthe crocata L.	Hemlock water dropwort	Apiaceae	Spain	A poultice is prepared with its tuberous roots and mixed with garlic, salt, vinegar, and olive oil.	(Gonzalez and Vallejo 2013)
80	Olea europaea L.	Common Olive	Oleaceae	Spain	Drink oil as an antidote	(Gonzalez and Vallejo 2013)
81	Olea ferruginea Wall. Ex Aitch.	Indian olive	Oleaceae	Pakistan	Roots are used	(Nasim et al. 2013)
82	Oxalis corniculate L.	Wood sorrel	Oxalidaceae	Pakistan	Leaves are used	(Nasim et al. 2013)
83	Phyla nodiflora (L.) Greene	Bukkan	Verbenaceae	Pakistan	Leaves are used	(Nasim et al. 2013)
84	Pinus roxburghii Sarg.	Chir	Pinaceae	Northern Pakistan	Oil is extracted from the wood and is rubbed on the infected area thrice a day.	(Butt et al. 2015; Gonzalez and Vallejo 2013)
85	Piper longum L.	long pepper	Piperaceae	India	Fruit is used to treat scorpion stings.	(Nasim et al. 2013)
86	Pistacia chinensis Bunge	Kangar	Anacardiaceae	Northern Pakistan	The powder of dried galls on the infected site is applied	(Butt et al. 2015)
87	Pistacia integerrima	Sweer	Anacardiaceae	Northern Pakistan	The powder obtained from dried galls and applied on wounds	(Butt et al. 2015)
88	Pistacia integrimma J. L. Stewart	Crab's claw	Anacardiaceae	Pakistan	Galls and bark are used	(Nasim et al. 2013)

#	Scientific name	Common name	Family	Location	Usage	Reference
89	*Platanus orientalis* L.	Chinar	Platanaceae	Northern Pakistan	2–3 teaspoons of fresh bark juice are given orally thrice daily, while 10–12 drops are used in the infected area.	(Butt et al. 2015)
90	*Prangospabularia* Lindl.	Cachryspabularia	Umbelliferae	Pakistan	Leaves are used	(Nasim et al. 2013)
91	*Prosopis cineraria* (L.) Druce	Spunge tree	Mimosaceae	Pakistan	Bark is used	(Nasim et al. 2013)
92	*Ricinus communis* L.	Erand	Euphorbiceae	Northern Pakistan, Andhra Pradesh, India	Fresh leaves are meshed, made into a paste, and applied to the wound.	(Butt et al. 2015; Basha and Murthy 2017)
93	*Rosmarinus officinalis* L.	Rosemary	Lamiaceae	Spain	The smoke from the combustion of rosemary is used.	(Gonzalez and Vallejo 2013)
94	*Rubia cordifolia* L.	Manjit	Rubiaceae	Northern Pakistan	Juice is extracted from different parts of the plant. Three spoons of the fluid are taken orally twice a day. It is also applied to wounds directly.	(Butt et al. 2015)
95	*Rutamontana* Mill.	Mountain Rue	Rutaceae	Spain	The poultice of the crushed fresh plant is applied. The plant is boiled with the liquid wet hot packs on the affected area.	(Gonzalez and Vallejo 2013)
96	*Sambucus nigra* L.	Elderberry	Caprifoliaceae	Spain	The smoke from the burning of older people is used.	(Gonzalez and Vallejo 2013)

(continued)

TABLE 5.1 (Continued)
A list of herbs that have been used historically to treat scorpion stings

Sl. No	Scientific Name	Common Name	Family	Country	Traditional Route of Administration	Reference(s)
97	*Sapindusmukorossi* Gaertn.	Raitha	Sapindales	Northern Pakistan	Fresh fruits are crushed, made into a paste, and applied for three days.	(Butt et al. 2015)
98	*Sarcostemmaviminale* (L.) R.Br.	Cynanchumviminale	Asclepiadaceae	Pakistan	Roots are used	(Nasim et al. 2013)
99	*Scilla hyacinthiana* (Roth.) Macbr.	Jangali Kanda	Liliaceae	Maharashtra, India	Bulb paste is applied on scorpion sting.	(Satpute 2019)
100	*Senecio jacobaea* D. Don	Common Ragwort	Asteraceae	Spain	Apply the poultices of the leaves.	(Gonzalez and Vallejo 2013)
101	*Sesamum indicum* L.	Sesame	Pedaliaceae	Saudi Arabia	Seed oil is applied to the sting area every 15 minutes.	(Fakhry, Migahid, and Anazi 2017)
102	*Skimmialaureola* (DC.) Decne.	Nair pat	Rutaceae	Pakistan	The whole plant is used.	(Nasim et al. 2013)
103	*Solanum tuberosum* L.	Irish potato	Solanaceae		A raw cut of potato is placed on the site of the sting.	(Gonzalez and Vallejo 2013)
104	*Strychnospotatorum* L.f.	Chilaginja, Chilla	Loganiaceae	Andhra Pradesh, India	Seeds are pasted by rubbing them on a rock and applying them.	(Krishna, Saidulu, and Hindumathi 2014)
105	*Tephrosia purpurea* (L.) Pers.	Vempali	Fabaceae	Andhra Pradesh, India	Leaf paste is applied over the sting, exposing the area to heat.	(Krishna, Saidulu, and Hindumathi 2014)
106	*Terminalia arjuna* (Roxb. Ex DC.) Wight & Arn	Maddi	Combretaceae	Andhra Pradesh, India	The aerial part of the plant is burned, and its ash is put on the sting site thrice a day.	(Basha and Murthy 2017)

107	*Trachydiumroylei* Lindl.	*Churoo*	Umbelliferae	Pakistan	Leaves are used.	(Nasim et al. 2013)
108	*Tribulus terrestris* L.	Gokharu, Sarata	Zygophyllaceae	Maharashtra, India	50 mL root extract is given twice daily for up to 3 days.	(Satpute 2019)
109	*Urginea maritima* (L.) Baker	Red squill	Liliaceae	Spain	The bulb is directly used by rubbing it on the sting area.	(Gonzalez and Vallejo 2013)
110	*Valerianajatamansi* D. Don	Indian valerian	Caprifoliaceae	Northern Pakistan	The dried root powder is mixed with honey and rubbed on the wound as needed.	(Butt et al. 2015)
111	*Verbena officinalis* L.	Shomakai	Verbenaceae	Northern Pakistan	The root paste is used as an external application that helps wounds heal.	(Butt et al. 2015)
112	*Vicia faba* L.	Jaba	Fabaceae	Spain	Halved dried seed is placed on the area of the sting.	(Gonzalez and Vallejo 2013)

The data is reported from 2013 to 2022 (Table and legends were adapted with permission from Nath and Mukherjee 2023; License Number:5692371089491).

the plant extracts are generic and less efficient as they do not specifically target a particular type of scorpion. Third, there is little to no knowledge of the mechanism of action of plant extracts or even isolated pure compounds, which means that they are unsuitable for use in modern therapeutic settings. Thus, uncovering a treasure trove of potential phytomedicine against scorpionism requires the scientific validation of the plants and extracts used in conventional therapy.

5.10   CONCLUSION

It has been shown that independent of the treatment—such as antivenom or intensive care—mortality decreases in areas where health officials have anticipated and coordinated the handling of scorpion stings (Chippaux and Goyffon 2008). However, deciding between the two possibilities necessitates a realistic evaluation of all pertinent factors, including logistics. Additionally, one should consider the ease of care and speed of recovery; mortality should not be the primary objective. Selecting symptomatic treatments can be challenging due to the complexity of contradictory clinical symptoms, especially considering the rapid symptom development and the potential for sequelae to emerge. While some medicines can be taken regularly with proper and precise regimens, many others still fall under the authority of the specialist.

In contrast, antivenom therapy, which rapidly eliminates venom from the organism, is now more straightforward to administer and carries fewer hazards. Even when aided by others who have received training, fragments of pure IgG are highly safe and effective. When administered early, they can limit transfers to referral hospitals and avoid significant problems, but this approach necessitates easy availability of antivenom in remote healthcare facilities. Immunotherapy in conjunction with symptomatic treatments is still frequently required. These therapeutic approaches complement one another and provide a satisfying response to most problems that develop in tropical nations (Bawaskar and Bawaskar 2011). All countries with a high incidence of scorpion stings must educate their health experts on this combined therapy. The public must be informed about these novel therapy approaches to encourage early referral to medical facilities.

REFERENCES

Abbas AK, and Litchman AH. 2003. Immediate hypersensitivity. In *Cellular and Molecular Immunology*, 5th ed., edited by AK Abbas and AH Litchman. Philadelphia: Saunders Elsevier Science. ISBN: 9780323757485.

Abbas AK, Lichtman AH, and Pillai S. 2018. Chapter 18: Immunity to tumors. In *Cellular Molecular Immunology*, edited by P. Ninth. Philadelphia: Elsevier:397–416.

Abroug F, ElAtrous S, Nouria S, Haguiga H, Touzi N, and Bouchoucha S. 1999. Serotherapy in scorpion envenomation: a randomised controlled trial. *The Lancet* 354 (9182):906–909.

Al-Asmari AK, Al-Seif AA, Hassen MA, and Abdulmaksood NA. 2008. Role of prazosin on cardiovascular manifestations and pulmonary oedema following severe scorpion stings in Saudi Arabia. *Saudi Medical Journal* 29 (2):299–302.

Amann R, and Peskar BA. 2002. Anti-inflammatory effects of aspirin and sodium salicylate. *European Journal of Pharmacology* 447 (1):1–9.

Arrowsmith JB, Kennedy DL, Kuritsky JN, and Faich GA. 1987. National patterns of aspirin use and Reye syndrome reporting, United States, 1980 to 1985. *Pediatrics* 79 (6):858–863.

Bahekar S, Kale R, and Nagpure S. 2012. A review on medicinal plants used in scorpion bite treatment in India. *Mintage Journal of Pharmaceutical and Medical Sciences* 1 (1):1–6.

Basha SKM, and Murthy NCV. 2017. Antidotes used for scorpion sting by the tribals of Siddeswarm sacred grooves of SPSR Nellore DT. AP. *International Journal of Engineering Research & Technology* 4 (5):65–67.

Bawaskar HS, and Bawaskar PH. 1986. Prazosin in management of cardiovascular manifestations of scorpion sting. *The Lancet* 327 (8479):510–511.

Bawaskar HS, and Bawaskar PH. 1991. Treatment of cardiovascular manifestations of human scorpion envenoming: is serotherapy essential? *The Journal of Tropical Medicine Hygiene* 94 (3):156–158.

Bawaskar HS, and Bawaskar PH. 1992. Management of the cardiovascular manifestations of poisoning by the Indian red scorpion (*Mesobuthus tamulus*). *Heart* 68 (11):478–480.

Bawaskar HS, and Bawaskar PH. 1994. Vasodilators: scorpion envenoming and the heart (an Indian experience). *Toxicon* 32 (9):1031–1040.

Bawaskar HS, and Bawaskar PH. 1996. Severe envenoming by the Indian red scorpion *Mesobuthus tamulus*: the use of prazosin therapy. *QJM: An International Journal of Medicine* 89 (9):701–704.

Bawaskar HS, and Bawaskar PH. 2000. Prazosin therapy and scorpion envenomation. *The Journal of the Association of Physicians of India* 48 (12):1175–1180.

Bawaskar HS, and Bawaskar PH. 2007. Utility of scorpion antivenin vs prazosin in the management of severe *Mesobuthus tamulus* (Indian red scorpion) envenoming at rural setting. *Journal of the Association of Physicians of India* 55:14–21.

Bawaskar HS, and Bawaskar PH. 2011. Efficacy and safety of scorpion antivenom plus prazosin compared with prazosin alone for venomous scorpion (*Mesobuthus tamulus*) sting: randomised open label clinical trial. *British Medical Journal* 342:c7136.

Biaggioni I, and Robertson D. 2012. Adrenoceptor agonists & sympathomimetic drugs. In *Basic & Clinical Pharmacology*, 12th ed., edited by BG Katzung, SB Masters, and AJ Trevor. New York: The McGraw-Hill Companies.

Binorkar SV. 2012. Herbal medicines used in the management of scorpion sting in traditional practices—a review. *American Journal of PharmTech Research* 2 (3):243–256.

Bosnak M, Levent YH, Ece A, Yildizdas D, Yolbas I, Kocamaz H, Kaplan M, and Bosnak V. 2009. Severe scorpion envenomation in children: management in pediatric intensive care unit. *Human Experimental Toxicology* 28 (11):721–728.

Boyer LV, Theodorou AA, Berg RA, Mallie J, Chávez-Méndez A, García-Ubbelohde W, Hardiman S, and Alagón A. 2009. Antivenom for critically ill children with neurotoxicity from scorpion stings. *New England Journal of Medicine* 360 (20):2090–2098.

Bullock S, and Manias E. 2013. *Fundamentals of Pharmacology*. Pearson Higher Education AU. ISBN: 1442564415

Butt MA, Ahmad M, Fatima A, Sultana S, Zafar M, Yaseen G, Ashraf MA, Shinwari ZK, and Kayani S. 2015. Ethnomedicinal uses of plants for the treatment of snake and scorpion bite in Northern Pakistan. *Journal of Ethnopharmacology* 168:164–181.

Carmo AO, Chatzaki M, Horta CC, Magalhães BF, Oliveira-Mendes BB, Chávez-Olórtegui C, and Kalapothakis E. 2015. Evolution of alternative methodologies of scorpion antivenoms production. *Toxicon* 97:64–74.

Celis A, Gaxiola-Robles R, Sevilla-Godínez E, and J Armas. 2007. Trends in mortality from scorpion stings in Mexico, 1979–2003. *Revista Panamericana de Salud Publica= Pan American Journal of Public Health* 21 (6):373–380.

Chippaux JP, and Goyffon M. 1998. Venoms, antivenoms and immunotherapy. *Toxicon* 36 (6):823–846.

Chippaux JP, and Goyffon M. 2008. Epidemiology of scorpionism: a global appraisal. *Acta Tropica* 107 (2):71–79.

Chippaux JP, Stock RP, and Massougbodji A. 2010. Methodology of clinical studies dealing with the treatment of envenomation. *Toxicon* 55 (7):1195–1212.

Chrousos G. 2012. Adrenocorticosteroids & adrenocortical antagonists. In *Basic & Clinical Pharmacology*, 12th ed., edited by B Katzung, SB Masters, and AJ Trevor. New York: McGraw-Hill.

Cruce J, and Lewis R. 2004. Types I, II, III, and IV hypersensitivity. In Atlas Immunol, 2nd ed. Florida: CRC Press.

Cupo P. 2015. Clinical update on scorpion envenoming. *Revista da Sociedade Brasileira de Medicina Tropical* 48:642–649.

Cupo P, and Hering SE. 2002. Cardiac troponin I release after severe scorpion envenoming by *Tityus serrulatus*. *Toxicon* 40 (6):823–830.

Das B, Madhubala D, Mahanta S, Patra A, Puzari U, Khan MR, and Mukherjee AK. 2023. A novel therapeutic formulation for the improved treatment of Indian red scorpion (*Mesobuthus tamulus*) venom-induced toxicity-tested in *Caenorhabditis elegans* and Rodent Models. *Toxins* 15 (8):504.

Das B, Patra A, Puzari U, Deb P, and Mukherjee AK. 2022. In vitro laboratory analyses of commercial anti-scorpion (*Mesobuthus tamulus*) antivenoms reveal their quality and safety but the prevalence of a low proportion of venom-specific antibodies. *Toxicon* 215:37–48.

de Dàvila CAM, Dàvila DF, Donis JH, Arata de Bellabarba G, Villarreal V, and Barboza JS. 2002. Sympathetic nervous system activation, antivenin administration and cardiovascular manifestations of scorpion envenomation. *Toxicon* 40 (9):1339–1346.

Dehesa-Dávila M, and Possani LD. 1994. Scorpionism and serotherapy in Mexico. *Toxicon* 32 (9):1015–1018.

Dehghani R, and Arani MG. 2015. Scorpion sting prevention and treatment in ancient Iran. *Journal Traditional Complement Medicine* 5 (2):75–80.

De Rezende NA, Borges Dias M, Campolina D, Chavez-Olortegui C, Rebeiro Diniz C, and CF Amaral. 1995. Efficacy of antivenom therapy for neutralizing circulating venom antigens in patients stung by *Tityus serrulatus* scorpions. *The American Journal of Tropical Medicine Hygiene* 52 (3):277–280.

Deshpande SB, Pandey R, and Tiwari AK. 2008. Pathophysiological approach to the management of scorpion envenomation. *Indian Journal of Physiology Pharmacology* 52 (3):311–314.

de Silva HA, Pathmeswaran A, Ranasinha CD, Jayamanne S, Samarakoon SB, Hittharage A, Kalupahana R, Ratnatilaka GA, Uluwatthage W, Aronson JK, and Armitage JM. 2011. Low-dose adrenaline, promethazine, and hydrocortisone in the prevention of acute adverse reactions to antivenom following snakebite: a randomised, double-blind, placebo-controlled trial. *PLoS Medicine* 8 (5):e1000435.

de Silva HA, Ryan NM, and Janaka de Silva, H. 2016. Adverse reactions to snake antivenom, and their prevention and treatment. *British Journal of Clinical Pharmacology* 81 (3):446–452.

Dnyaneshwar DS, and Sonali C. 2016. A critical review on anti-scorpion activity of herbs. *International Journal of Ayurveda Research* 4 (11):3354–3360.

Erridge C, Bennett-Guerrero E, and Poxton Ian R. 2002. Structure and function of lipopolysaccharides. *Microbes Infection* 4 (8):837–851.

Fakhry AM, Migahid MA, and Anazi HK. 2017. Herbal remedies used in the treatment of scorpion stings in Saudi Arabia. *Global Journal of Medicinal Plant Research* 5:1–8.

Freire-Maia L, and Campos JA. 1989. Pathophysiology and treatment of scorpion poisoning. In *Natural Toxins, Characterization, Pharmacology and Therapeutics. Proceedings of the 9th World Congress on Animal, Plant and Microbial Toxins*, edited by CL Ownby and GV Odell. Oxford, UK: Pergamon Press.

Freire-Maia L, Campos JA, and Amaral CFS. 1994. Approaches to the treatment of scorpion envenoming. *Toxicon* 32 (9):1009–1014.

Ghalim N, El-Hafny B, Sebti F, Heikel J, Lazar N, Moustanir R, and Benslimane A. 2000. Scorpion envenomation and serotherapy in Morocco. *The American Journal of Tropical Medicine Hygiene* 62 (2):277–283.

Gibly R, Williams M, Walter FG, McNally J, Conroy C, and Berg RA. 1999. Continuous intravenous midazolam infusion for *Centruroides exilicauda* scorpion envenomation. *Annals of Emergency Medicine* 34 (5):620–625.

Gonzalez JA, and Vallejo JR. 2013. The scorpion in Spanish folk medicine: a review of traditional remedies for stings and its use as a therapeutic resource. *Journal of Ethnopharmacology* 146 (1):62–74.

Goyffon M, Vachon M, and Broglio N. 1982. Epidemiological and clinical characteristics of the scorpion envenomation in Tunisia. *Toxicon* 20 (1):337–344.

Gupta SD, Debnath A, Saha A, Giri B, Tripathi G, Vedasiromoni JR, Gomes A, and Gomes A. 2007. Indian black scorpion (*Heterometrus bengalensis Koch*) venom induced antiproliferative and apoptogenic activity against human leukemic cell lines U937 and K562. *Leukemia Research* 31 (6):817–825.

Gupta V. 2006. Prazosin: a pharmacological antidote for scorpion envenomation. *Journal of Tropical Pediatrics* 52 (2):150–151.

Gutiérrez JM, Solano G, Pla D, Herrera M, Segura Á, Vargas M, Villalta M, Sánchez A, Sanz L, Lomonte B, and León G. 2017. Preclinical evaluation of the efficacy of antivenoms for snakebite envenoming: state-of-the-art and challenges ahead. *Toxins* 9 (5):163.

Hammoudi-Triki D, Ferquel E, Robbe-Vincent A, Bon C, Choumet V, and Laraba-Djebari F. 2004. Epidemiological data, clinical admission gradation and biological quantification by ELISA of scorpion envenomations in Algeria: effect of immunotherapy. *Transactions of the Royal Society of Tropical Medicine Hygiene* 98 (4):240–250.

Ismail, M. 1994. The treatment of the scorpion envenoming syndrome: the Saudi experience with serotherapy. *Toxicon* 32 (9):1019–1026.

Ismail, M. 1995. The scorpion envenoming syndrome. *Toxicon* 33 (7):825–858.

Katzung BG, and Julius DJ. 2001. Histamine, serotonin, and the ergot alkaloids. *Basic Clinical Pharmacology* 13:437–68.

Koseoglu Z, and Koseoglu A. 2006. Use of prazosin in the treatment of scorpion envenomation. *American Journal of Therapeutics* 13 (3):285–287.

Krishna NR, Saidulu C, and Hindumathi A. 2014. Ethnomedicinal uses of some plant species by tribal healers in Adilabad district of Telangana state, India. *World Journal of Pharmaceutical Research* 3:545–561.

Krishnan A, Sonawane RV, and Karnad DR. 2007. Captopril in the treatment of cardiovascular manifestations of Indian red scorpion (*Mesobuthus tamulus concanesis Pocock*) envenomation. *Journal of the Association of Physicians of India* 55 (1):22–26.

Latifi M, and Tabatabai M. 1979. Immunological studies on Iranian scorpion venom and antiserum. *Toxicon* 17 (6):617–620.

León G, Herrera M, Segura Á, Villalta M, Vargas M, and Gutiérrez JM. 2013. Pathogenic mechanisms underlying adverse reactions induced by intravenous administration of snake antivenoms. *Toxicon* 76:63–76.

Morais V. 2018. Antivenom therapy: efficacy of premedication for the prevention of adverse reactions. *Journal of Venomous Animals Toxins Including Tropical Diseases* 24:7. DOI: 10.1186/s40409-018-0144-0

Morais V, Berasain P, Ifrán S, Carreira S, Tortorella M, Negrín A, and Massaldi H. 2012. Humoral immune responses to venom and antivenom of patients bitten by Bothrops snakes. *Toxicon* 59 (2):315–319.

Morais VM, and Massaldi H. 2009. Snake antivenoms: adverse reactions and production technology. *Journal of Venomous Animals Toxins Including Tropical Diseases* 15:2–18.

MukherjeeAK. 2021. *'Big Four' Snakes of India*. Singapore: Springer. ISBN: 978-981-16-2895-5

Mukherjee AK, and Chattopadhyay DJ. 2022. Potential clinical applications of phytopharmaceuticals for the in-patient management of coagulopathies in COVID-19. *Phytotherapy Research* 36 (5):1884–1913.

Nascimento Jr, EB, Costa KA, Bertollo CM, Oliveira ACP, Rocha LT, Souza AL, Glória MBA, Moraes-Santos T, and Coelho MM. 2005. Pharmacological investigation of the nociceptive response and edema induced by venom of the scorpion *Tityus serrulatus*. *Toxicon* 45(5):585–593.

Nasim MJ, Asad MH, Sajjad A, Khan SA, Mumtaz A, Farzana K, Rashid Z, and Murtaza G. 2013. Combating of scorpion bite with Pakistani medicinal plants having ethno-botanical evidences as antidote. *Acta Poloniae Pharmaceutica—Drug Research* 70:387–394.

Nath S, and Mukherjee AK. 2022. Ethnomedicines for the treatment of scorpion stings: a perspective study. *Journal of Ethnopharmacology* 305:116078.

Otero-Patiño R, Cardoso JL, Higashi HG, Nunez V, Diaz A, Toro MF, Garcia ME, Sierra A, Garcia LF, Moreno AM, and Medina MC. 1998. A randomized, blinded, comparative trial of one pepsin-digested and two whole IgG antivenoms for Bothrops snake bites in Uraba, Colombia. The Regional Group on Antivenom Therapy Research (REGATHER). *The American Journal of Tropical Medicine Hygiene* 58 (2):183–189.

Ozkan O, and Yağmur EA. 2017. Neutralization capacity of monovalent antivenom against existing lethal scorpions in the Turkish Scorpiofauna. *Iranian Journal of Pharmaceutical Research* 16 (2):653.

Peker E, Oktar S, Dogan M, Kaya E, and Duru M. 2010. Prazosin treatment in the management of scorpion envenomation. *Human Experimental Toxicology* 29 (3):231–233.

Prasad R, Kumar A, Jain D, Das BK, Singh UK, and Singh TB. 2020. Echocardiography versus cardiac biomarkers for myocardial dysfunction in children with scorpion envenomation: An observational study from tertiary care center in northern India. *Indian Heart Journal* 72 (5):431–434.

Prasad R, Mishra OP, Pandey N, and Singh TB. 2011. Scorpion sting envenomation in children: factors affecting the outcome. *The Indian Journal of Pediatrics* 78:544–548.

Puzari U, Fernandes PA, and Mukherjee AK. 2022. Pharmacological re-assessment of traditional medicinal plants-derived inhibitors as antidotes against snakebite envenoming: a critical review. *Journal of Ethnopharmacology* 292:115208.

Satpute SV. 2019. Medicinal plants used in the traditional treatment of scorpion stings and mad-dog bite. *International Journal of Current Research* 11 (3):1866–1868.

Seifert SA. 2001. Atropine use in centruroides scorpion envenomation—contraindicated or not? (Commentary). *Journal of Toxicology: Clinical Toxicology* 39 (6):599–599.

Sevcik C, D'suze G, Díaz P, Salazar V, Hidalgo C, Azpúrua H, and Bracho N. 2004. Modelling *Tityus* scorpion venom and antivenom pharmacokinetics. Evidence of active immunoglobulin G's F (ab')$_2$ extrusion mechanism from blood to tissues. *Toxicon* 44 (7):731–741.

Silawal S, Kohl B, Shi J, and Schulze-Tanzil, G. 2021. Complement regulation in human tenocytes under the influence of Anaphylatoxin C5a. *International Journal of Molecular Sciences* 22 (6):3105.

Silva LT, Junior RS, Teixeira de Carvalho TX, Moutinho Pataca LC, and Guilherme Dias Heneine L. 2023. Analysis of antibodies avidity for *Tityus serrulatus* scorpion venom in antivenom production and its potential for application as a potency test. *Toxicon* 236:107315.

Silverstein AM. 2000. Clemens Freiherr von Pirquet: explaining immune complex disease in 1906. *Nature Immunology* 1 (6):453–455.

Sofer S, Shahak E, and Gueron M. 1994. Scorpion envenomation and antivenom therapy. *The Journal of Pediatrics* 124 (6):973–978.

Soulaymani-Bencheikh R, Faraj Z, Semlali I, Khattabi A, Skalli S, Benkirane R, and Badri M. 2002. Epidemiological aspects of scorpion stings in Morocco. *Revue D'epidemiologie et de Sante Publique* 50 (4):341–347.

Soulaymani Bencheikh R, Idrissi M, Tamim O, Semlali I, Mokhtari A, Tayebi M, and Soulaymani A. 2007. Scorpion stings in one province of Morocco: epidemiological, clinical and prognosis aspects. *Journal of Venomous Animals Toxins Including Tropical Diseases* 13:462–471.

Squaiella-Baptistao CC, Roberto Marcelino J, Eduardo Ribeiro da Cunha L, María Gutiérrez J, and Tambourgi DV. 2014. Anticomplementary activity of horse IgG and F (ab') $_2$ antivenoms. *The American Journal of Tropical Medicine Hygiene* 90 (3):574.

Stone SF, Geoffrey KI, Shahmy S, Mohamed F, Abeysinghe C, Karunathilake H, Ariaratnam A, Jacoby-Alner TE, Cotterell CL, and Brown SG. 2013. Immune response to snake envenoming and treatment with antivenom; complement activation, cytokine production and mast cell degranulation. *PLoS Neglected Tropical Diseases* 7 (7):e2326.

Traynor K. 2011. Scorpion antivenin approved. *American Journal of Health-System Pharmacy* 68 (18):1668–1669.

Venancio EJ, Portaro FC, Kuniyoshi AK, Carvalho DC, Pidde-Queiroz G, and Tambourgi DV. 2013. Enzymatic properties of venoms from Brazilian scorpions of *Tityus* genus and the neutralisation potential of therapeutical antivenoms. *Toxicon* 1(69):180–90.

Warrell DA. 1997. Prazosin: scorpion envenoming and the cardiovascular system. *Tropical Doctor* 27(1):1–1.

Warrell DA. 2019. Venomous bites, stings, and poisoning: an update. *Infectious Disease Clinics* 33 (1):17–38.

WHO. 2019. Snakebite envenoming: a strategy for prevention and control. ISBN: 978 92 4 151564 1

Williams DJ, Habib AG, and Warrell DA. 2018. Clinical studies of the effectiveness and safety of antivenoms. *Toxicon* 150:1–10.

6 Scorpion Venom Toxins
Biomedical and Therapeutic Applications

6.1 SCORPIONS: A VAST TREASURY OF LIVING RESOURCES

Venom is a potent mixture of bioactive elements with distinct pharmacological characteristics that has led to the discovery of numerous drug prototypes. Due to the existence of some toxic peptides, scorpion venom is like a two-edged sword, in that it can cause major clinical manifestations after a sting while also having some whimsical medicinal, biomedical, and industrial uses that improve societal health. They might offer a viable framework for the creation of novel medications. Scorpions are found all over the world, and those that are significant medically are members of the Buthidae family (whose stings are also fatal), except *Hemiscorpius lepturus* (Hemiscorpiidae) and *Pandinus imperator* (Scorpionidae) scorpions, both of which are equally dangerous.

6.2 BIOMEDICAL APPLICATION OF SCORPION VENOM TOXINS

For many years, scorpion toxins have been referenced by Ayurvedic, Unani, Chinese, and homoeopathic medical systems(Wang et al. 1991). In addition to the numerous proteins and peptides that are contained in scorpion venoms and which have a variety of therapeutic benefits, technological advancements have made it possible to isolate and purify the peptides from scorpion venom for use as therapeutic or medicinal agents (Joseph and George 2012). Because of its cytotoxic, apoptogenic, immunosuppressive, and antiproliferative effects, scorpion venom, a mixture of diverse components including protein, peptide, biogenic agents, mucoproteins, organic salts, and neurotoxin, has demonstrated a potential therapeutic application (Ahmadi et al. 2020).

6.2.1 IN TREATMENT OF CANCER

According to studies on using scorpion venom in medicine, its toxins may be a feasible alternative to cancer treatments (Desales-Salazar et al. 2020; Mishal et al. 2013). Voltage-gated Na^+ channel development is crucial for creating novel drug prototypes (Hmed, Serria, and Mounir 2013). These channels play a role in metastasis growth after cancer treatment as they are present in many malignancies. For example, K^+ channels also affect cancer cell growth (Villalonga et al. 2007).

 DOI: 10.1201/9781003540816-6

Despite their significant contribution, current cancer treatments like radiotherapy and chemotherapy do not provide satisfactory results, and they also have adverse side effects on the healthy tissue surrounding the tumor (DeSantis et al. 2014; Rao et al. 2015). Ion transport through channels across the cell membrane controls the motility and survival of cancer cells. Ion transport across the cell membrane maintains the essential functions of tumor cells, including migration, cell cycle progression, proliferation, and cell volume regulation (Turner and Sontheimer 2014).

Various natural products or synthetic analogs are prescribed clinically for cancer treatment (Yu and Meng 2016). Between 1981 and 2010, the US Food and Drug Administration (FDA) approved 98 novel anticancer medications, of which 78 were natural products or derived from raw materials, and just 20 were synthetic pharmaceuticals (Newman and Cragg 2012). Animal-derived compounds (mainly those from arthropods) are rarely utilized as therapeutic prototypes or in clinical practice and trials (Rapôso 2017).

Scorpion venoms in in vitro conditions have been demonstrated to inhibit the growth of several cancer cells, for example, human leukemia, prostate cancer, and neuroblastoma (Gupta et al. 2007; Zhang et al. 2009; Zargan et al. 2011). Scorpion venom peptides exhibit significant effects on cancer cells, including four potential mechanisms: (i) initiation of cell cycle arrest, growth inhibition, and apoptosis; (ii) angiogenesis inhibition; (iii) invasion and metastasis inhibition; and (iv) specific transmembrane channels blocking (Rapôso 2017). Generally, they perform anticancer characteristics through different mechanisms such as blocking a particular channel of ion (Jager et al. 2004) or binding to a specific site of cancer cells, thus inhibiting their invasion in the plasma membrane (Deshane, Garner, and Sontheimer 2003), or activation of intracellular pathways and inducing apoptosis (Gupta et al. 2010), leading to impairment of multiple hallmarks of different types of cancer in vitro and in vivo such as neuroblastoma-, glioma-, leukemia-, lymphoma-, lung-, breast-, pancreatic-, hepatoma-, prostate-, and other models of cancer (Rapôso 2017). The first scorpion venom reported to have anticancer effects was obtained from the Chinese scorpion *Buthus martensii Karsch* (Heinen and Veiga 2011).

Table 6.1 shows different scorpion venom peptides with anticancer properties. The venom of *Rhopalurus junceus* has also been used in in vivo toxicological studies; however, toxic effects have not been observed whether taken orally (2,000 mg/kg) or intraperitoneally (10 mg/kg) (García-Gómez et al. 2011; Lagarto et al. 2020). Research conducted on the pharmacokinetics and biodistribution of venom given to mice with breast tumors orally or intravenously showed that the venom's medium residence time (MRT) in tumor tissue was more significant than in the other organs examined, suggesting a high selectivity over tumor tissue and boosting its antitumor effect (Díaz-García et al. 2013). Additionally, mice with breast tumors that received 10 intraperitoneal injections of *R. junceus* venom (3.2 mg/kg) demonstrated reduced tumor growth and decreased levels of the tumor markers Ki67 and CD31, demonstrating the venom's anticancer potential (Díaz-García et al. 2013). Even though they are two of the most venomous species of scorpion, *Leiurus quinquestriatus* (Al-Asmari, Islam, and Al-Zahrani 2016) and *Androctonus amoreuxi* (Salem et al. 2016) have been investigated, showing encouraging in vivo anticancer effects.

TABLE 6.1
Different scorpion venom peptides with anticancer properties

S. No.	Name of Scorpion	Isolated Peptides	Name of Anticancer Properties	References
1	*Buthus martensii Karsch*	PSEV, BmKn-2, LMWSVP, GST-BmKCT and Ad-BmKCT, Ragap, BmKKx2, TM-601	Human leukemia (K562); murine hepatoma; human lung (A549); human oral squamous carcinoma (HSC-4); human mouth epidermoid carcinoma (KB); human hepatoma (SMMC7721); rat glioma (C6); human anaplastic astrocytoma (SHG-44); rat glioma (C6); human myelogenous leukemia (K562); rat glioma (F98); human glioblastoma (U87).	(Rapôso 2017)
2	*Tityus serrulatus*	TiTx gamma, TsIV-5, TsAP-2, TsAP-1	Mouse neuroblastoma, human squamous carcinoma (NCIeH157); human lung adenocarcinoma (NCIeH838); human androgen-independent prostate adenocarcinoma (PC-3); human breast carcinoma (MCF-7); human glioblastoma (U251), human squamous carcinoma (NCIeH157); human lung adenocarcinoma (NCIeH838)	(Kirsch et al. 1989; Guo et al. 2013)
3	*Mesobuthus tamulus*	Iberiotoxin (IbTX)	Human glioma	(Ru et al. 2014)

6.2.2 IN TREATMENT OF CARDIOVASCULAR ABNORMALITIES

Peptides from scorpion venom play an active factor in treating cardiovascular abnormalities. Vasodilators in scorpion venom are responsible for the dilation of blood vessels and thus can decline high blood pressure. Toxins from *Androctonus australis* secrete some atrial natriuretic peptides. Similarly, Bmk I toxin isolated from *Buthus martensii Karsch* helps to maintain cardiac contraction (Lu et al. 2006; McLane, Joerger, and Mahmoud 2008). Again, a tripeptide KPP isolated from *Tityus serrulatus* scorpion venom properties in cardiovascular disease therapeutics proved to be a drug prototype for treating Chagas disease caused by the *Trypanosoma cruzi* parasite. Its chronic phase can be characterized by heart damage or chronic Chagas' cardiomyopathy (de Oliveira Pimentel et al. 2021).

6.2.3 IN THE TREATMENT OF HEMATOLOGICAL DISEASES

Scorpion venom intervenes in blood coagulation by accelerating or inhibiting the coagulation process. Toxins (polypeptides) from scorpion *Buthus martensii Karsch* show some anti-thrombotic action, which is resistant to blood platelet aggregation and raise the concentration of prostaglandin I2 in plasma (Song et al. 2005).

6.2.4 IN TREATMENT OF AUTOIMMUNE DISEASE

Autoimmune disease is the world's third leading mortality-causing disease after cancer and cardiovascular disorders. Its mechanism of pathophysiology is still unclear, although B- and T-lymphocytes play an essential role in the occurrence and development of this disease (Chatenoud 2016; Bach 2002). The anti-inflammatory and immunosuppressive agents, including nonsteroidal anti-inflammatory drugs, engineered biologics, glucocorticoids, monoclonal antibodies, and fusion proteins, are commonly used clinical medications for the treatment of autoimmune disorders that are selective for signaling pathways and specific immune cell subsets (Chatenoud 2016; Rose 2016).

As the effects of the medications may not be specific against the particular antigen, their long-term usage may lead to some side effects, such as renal and hepatic injuries, anaphylaxis, etc. It has been reported that in the treatment of autoimmune disease, scorpion venom peptides play a significant role (Shen et al. 2017) as scorpion venom has a vast resource of K^+ channel specific peptides, the kv1.3 channels (which comprise six membrane-spanning helical segments named S1–S6); on the other hand, they are the validated target as well as a new therapeutic target for the treatment of autoimmune disease (Chandy et al. 2004; Wang and Xiang 2013; Perez-Verdaguer et al. 2016).

A study in 1984 found that these channels are expressed in T-lymphocytes (Chandy et al. 1984; DeCoursey et al. 1984). This channel unit comprises four identical, independent, and non-covalently connected subunits. Each subunit is again composed of six transmembrane α-helices connected by intra- and extra-membranous loops. The positively charged S4 segment moves inward or outward depending on the potential of the cell membrane, which further drives the closing or opening of the Kv1.3 channel (Shen et al. 2017).

Some scorpion venom peptides are blockers of the K^+ channel and are used for treating rheumatoid arthritis, spondyloarthritis (Paksoy and Tetik 2022), bone resorption, multiple sclerosis, and other autoimmune diseases. Noxiustoxin is the first K^+ channel-blocking toxin isolated from scorpions, which was later successfully found in venomous scorpions (Carbone et al. 1982; Domingos Possani, Martin, and Svendsen 1982). Kaliotoxin (KTX) and OSK1 toxins isolated from the *Androctonus mauretanicus* and *Orthochirus scrobiculosus* scorpions, respectively, are immunosuppressive agents for the treatment of autoimmune disease (Chen and Chung 2012; Crest et al. 1992).

A clinical report was received from a 69-year-old patient admitted after being stung by *Mesobuthus eupeus* who had sacroiliitis symptoms for many years, and he recovered completely from this disorder post scorpion sting envenomation (Paksoy and Tetik 2022). The *M. eupeus* venom has been shown to have an immunomodulating effect; its immunosuppressant effect can find application in treating autoimmune diseases. A study investigating the results of the venom of the *M. eupeus* species on chickens concluded that it both increases and decreases the immune response due to the different protein structures it contains (Khosravi et al. 2017; Mirshafiey 2007). It is contemplated that changes in the humoral immune response can be specified by purifying protein structures separately to use this venom in autoimmune diseases (Zhao et al. 2015a). Besides, the KCa1.1 β1-3–specific venom peptide blocker, namely, iberiotoxin (IbTX) (synthetic), reduces disease severity in animal models with rheumatoid arthritis devoid of inducing significant side effects (Tanner et al. 2018).

6.2.5 In Treatment of Diabetes

Scorpion venom toxins are reportedly effective in treating diabetes and have properties of β-islets (Xie and Herbert 2012; Bouafir, Ait-Lounis, and Laraba-Djebari 2017). Toxins isolated from *Tityus bahiensis* and *T. serrulatus* venom enhance the proliferation of β cells in the pancreas (El-Ghlban et al. 2014; Bouafir, Ait-Lounis, and Laraba-Djebari 2017). Diabetic foot ulcer is one of the common complications of diabetes mellitus, which leads to foot infections, ulcers, and tissue destruction associated with distal extremity nerve abnormalities and peripheral vascular lesions. Whole body extract (telson containing venom gland discarded) from *Scorpio maurus palmatus* showed antidiabetic properties in alloxan-induced mice where islets of Langerhans of mice pancreas become improved after treatment with scorpion extract (Mohamed et al. 2019).

6.2.6 In Treatment of Erectile Dysfunction

Erectile dysfunction is a multifarious neurovascular phenomenon affecting men of all ages (Ia 1999). It is a persistent failure to sustain or accomplish a penile erection, and nitric oxide (NO) is the primary vasodilator for this event (Nunes and Webb 2012; Toda et al. 2005) (Figure 6.1).

Scorpions belonging to the Buthidae family, except Hemiscorpion, cause priapism. Venom toxins isolated from African scorpion *Leiurus quinquestriatus quinquestriatus*, scorpion *Buthus martensii Karsch*, *T. serrulatus* cause the release of NO in rats, rabbits, and humans, which, thus, relaxed their anococcygeus muscle (smooth muscle tissue of the urogenital tract), the smooth muscle tissue of urogenital track (Gwee, Cheah, and Gopalakrishnakone 1995; Srinivasan et al. 2001; Nunes et al. 2013) (Figure 6.2). Moreover, other scorpions, such as *A. australis* and *Buthotus judaicus*, have also been reported to cause relaxation of rabbit corpus cavernosum (Teixeira et al. 2001).

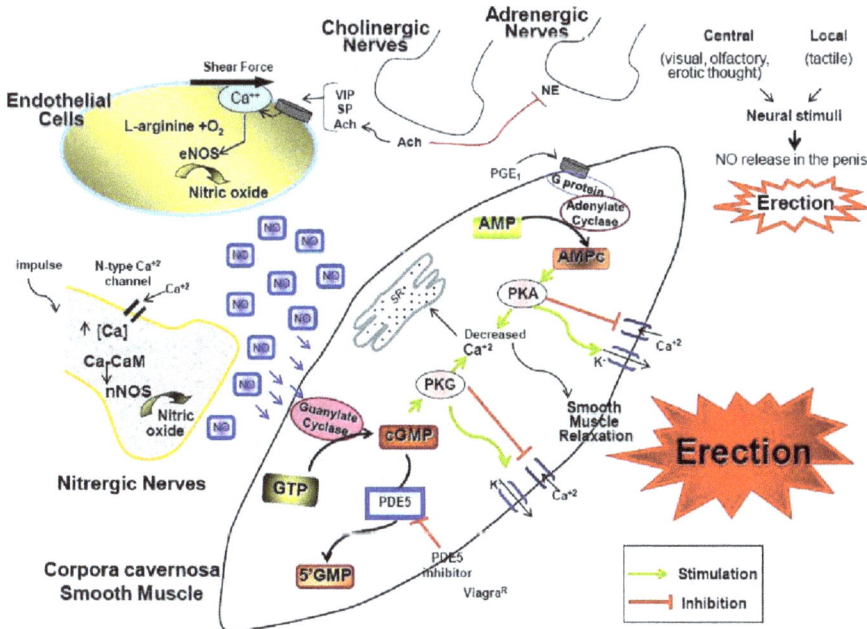

FIGURE 6.1 Mechanism of erectile dysfunction. (Figure and legend were adapted with permission from Nunes and Webb 2012; CC BY 3.0.)

6.2.7 ANTI-EPILEPTIC EFFECT OF SCORPION VENOM

Epilepsy is a non-contagious brain disease that affects millions of people of all ages globally, characterized by recurrent seizures and brief episodes of involuntary movements. Many anti-epileptic drugs are helpful for the treatment of patients with epileptic seizures. Still, due to some side effects, such as cognitive impairment, teratogenesis, chronic toxicity, and sedation, there is a high demand to improve the drug protocol (Raza et al. 2001). It has been reported that some toxins of scorpions have anti-epileptic effects, and recently α and β neurotoxins such as BmK IT2 and BmKAEP isolated from scorpion *Buthus martensii Karsch* have shown anti-epileptic properties in Sprague–Dawley rats (using coriaria lactone-induced model of epilepsy) (Figure 6.3) (Wang et al. 2020).

The scorpion venom heat-resistant peptide (SVHRP), which was extracted from the venom of *Buthus martensii Karsch,* has an anti-epileptic effect by decreasing seizure activity in the hippocampus of kainic acid-induced epileptic model rats (Chen et al. 2021). The anti-epileptic effects of the scorpion venom peptide HsTx2 were also observed. Pentylenetetrazol (PTZ)-induced epilepsy was decreased by the peptide in BALB/c mice by reducing astrocyte inflammation through the circ_0001293/miR-8114/TGF-2 axis. The findings highlighted the potential for future study into novel endogenous noncoding RNA-mediated processes of epilepsy as well as the application of exogenous peptide molecular probes as an unknown form of ASD (antiseizure medications) (Hu et al. 2022).

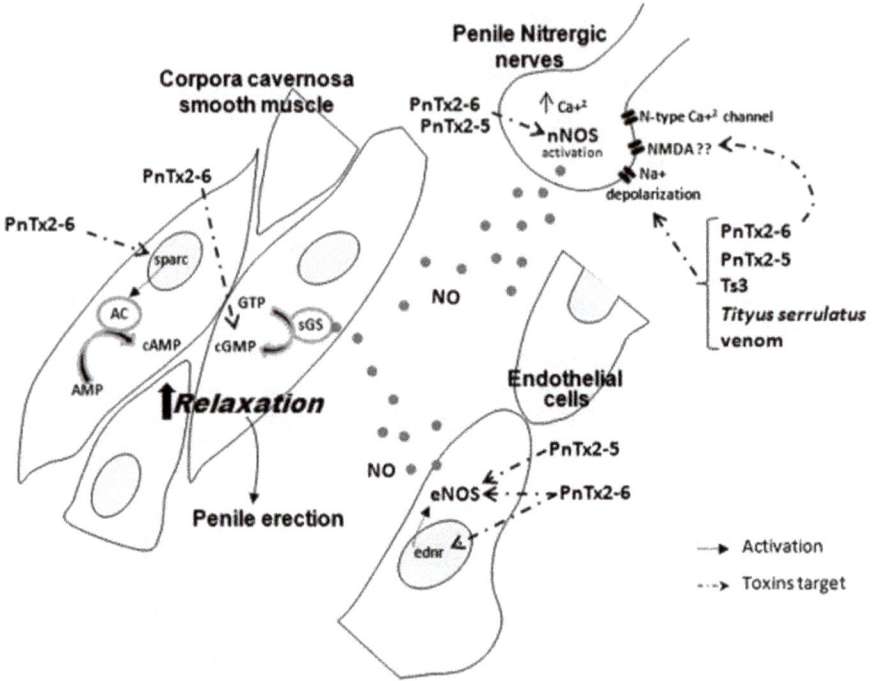

FIGURE 6.2 Scorpion toxins that potentiate erectile function. (Figure and legend were adapted and reproduced with permission from Nunes et al. 2013; License number: 5694261055129.)

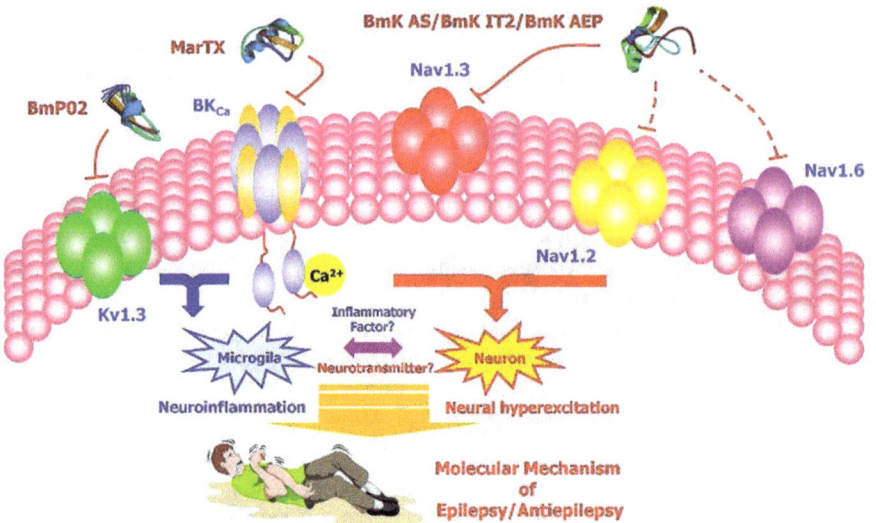

FIGURE 6.3 Anti-epilepsy effects of BmK venom toxins by targeting Na+/K+ channels .(Figure and legend were adapted with permission from Wang et al. 2020; CC BY 3.0.)

6.2.8 TREATMENT OF MICROBIAL INFECTIONS

Many scorpion venom-derived compounds have shown substantial antimicrobial activity in both in vitro and in vivo studies, as the growth of their infections can be inhibited by scorpion venom toxins due to their resistance. Venom peptides such as Bmkn2 and Ctriporin from *Mesobuthus martensii* and *Chaerilus tricostatus*, respectively, can specifically resist bacterial growth (Cao et al. 2014; Fan et al. 2011; Zeng et al. 2004). Antimicrobial peptides (AMPs) use different mechanisms of action to perform antibiotic properties; they disrupt cell membranes and leakage of pathogenic cells.

Different scorpion venom-derived AMPs are reported from the venom of other scorpion species, such as *A. australis*, *Parabuthus schlechteri*, *Opistophthalmus carinatus*, *P. imperator*, etc. Some scorpion venom peptides also show the property for treating viral infections (Lima et al. 2022). DENV and ZIKV isolated from *Scorpio maurus palmatus*, *Euscorpiops validus*, and *P. imperator* showed antiviral properties against dengue and Zika virus (Lima et al. 2022). Different peptides with antimicrobial activity have been isolated from scorpion venoms, such as mucroporin purified from *Lychas mucronatus* scorpion, showing antimicrobial properties against *Buthus subtilis* pathogen responsible for bacteremia, endocarditis, pneumonia, and septicemia (Dai et al. 2008). Peptides Bmkb1 and Bmkn2 isolated from *Buthus martensii Karsch* scorpion venom target against *Staphylococcus aureus* and *Escherichia coli* (Joseph and George 2012).

From *A. australis* venom, one compound, namely Androctonin (a peptide containing 25 amino acids with a molecular mass of 3.09 KDa), showed antimicrobial properties in multiple pathogens such as *Micrococcus luteus*, *A. viridians*, *Pseudomonas syringae*, *Salmonella typhimurium*, etc. (Hetru et al. 2000; Ehret-Sabatier et al. 1996). Again pandinin-1,-2, and scorpine are some of the peptides isolated from medically important scorpion *P. imperator* that target Gram-positive and Gram-negative bacteria, *B. subtilis*, *Klebsiella pneumonia*, and *Plasmodium berghei* pathogens (Conde et al. 2000; Harrison et al. 2014).

6.3 TREATMENT OF NEUROPATHIC PAIN: ROLE OF NA+ CHANNEL TOXINS

Neuropathic pain, a joint chronic pain, can happen when the somatosensory system is not working or damaged during trauma, injury to peripheral nerves (including nerve compression) or chemotherapy, and metabolic disorders such as diabetes (Aley and Levine 2002; Baron, Binder, and Wasner 2010). It is a continuous or episodic event associated with dysesthesia (unusual touch-based discomfort sensations) and pain without non-painful stimuli (Besson 1999; Loeser and Treede 2008). However, this disorder has not been characterized correctly due to a lack of precise molecular mechanisms (Attal et al. 2011; Doth et al. 2010; Langley et al. 2013). Neuropathic pain development has been linked to the activation of mGLUR5 receptors and N-methyl-D-aspartate receptor (NMDAR) by glutamate (Chen et al. 2019; Xie et al. 2017), activation of transcriptional factors such as inflammatory cytokine production (Kiguchi et al. 2012), and stimulation of the spinal cord's dorsal horn's nuclear factor-kappa B (NF-κB) cascade (Niederberger and Geisslinger 2008).

Furthermore, recent research revealed that peripheral and central immune cells, such as mast cells, T cells, astrocytes, microglia, and macrophages, may play a role in causing both peripheral and central sensitization (Inoue and Tsuda 2018). Some medicines' primary antinociceptive mechanism may be less microglial activation (Guida et al. 2012; Inyang et al. 2019). Furthermore, the factor involved in neuropathic pain may be the changes in ion channel and receptor expression levels, e.g., increased Na^+ channel expression on damaged neurons. Wallerian degeneration is linked to nerve growth factor release into the extracellular milieu surrounding spared fibers, which can activate higher expression levels of channels and receptors on healthy fibers (e.g., Na^+ channels, transient receptor potential (TRP) channels, or adrenoreceptors) (Kumar, Kaur, and Singh 2018).

As a new therapeutic drug prototype, different venom components have been used to treat neuropathic pain (Pennington, Czerwinski, and Norton 2018). As already described, scorpion venom is a complex cocktail of neurotoxins such as α and β toxins, which interact with voltage-gated Na^+ channels to fight against neuropathic pain (Cologna et al. 2009; Pucca et al. 2015; Pennington, Czerwinski, and Norton 2018). One component, namely TsNTxP, isolated from *T. serrulatus* interacts against a voltage-gated Na^+ channel, which increases membrane depolarization of peripheral nerves (Jonas et al. 1986) and leads to the glutamate release from synaptosomes (Massensini et al. 1998) associated interacting with Nav 1.2, Nav 1.4, and Nav 1.6 channels subtypes (Peigneur et al. 2015). Additionally, they inhibit Na^+ channel inactivation and encourage neurotransmitter release (Ghosh, Alajbegovic, and Gomes 2015; Gomez, Romano-Silva, and Prado 1995; Kirsch et al. 1989; Lima and Freire-Maia 1977). Thus, antinociceptive properties of TsNTxP were observed when examining the mouse nociception model against acute pain induced by intraplantar administration of capsaicin (Rigo et al. 2019). This component causes the intense sensation of red pepper. Previously, it was also reported to reduce the capsaicin-induced nociception by Na^+ channel blockers (Moon et al. 2012). TsNTxP's antinociceptive properties against acute pain led to an evaluation of the protein's effects in mouse neuropathy models (Rigo et al. 2019).

6.4   CONCLUSION

The scorpion venom comprises a broad spectrum of polypeptides with various bioactivities and high selectivity to identify multiple molecular pathways. These low-molecular-weight polypeptides are often compactly stabilized with disulfide bridges. Although it has been stated that they offer a great deal of potential as candidates for the creation of novel medications, more research is needed to start their clinical applications, which is still a hurdle. Public–private partnerships must be built and strengthened to progress the translational research movement in underdeveloped countries where these venoms are sourced and investigated. Also necessary are the implementation of public funds, the development of local enterprises, the establishment of facilities to expand translational research, and the encouragement of partnerships between academic institutions and the pharmaceutical industry. Numerous scorpion species still exist with undiscovered venoms, necessitating swift action to reveal their medical significance.

REFERENCES

Ahmadi S, Knerr JM, Argemi L, Bordon KC, Pucca MB, Cerni FA, Arantes EC, Çalışkan F, and Laustsen AH. 2020. Scorpion venom: detriments and benefits. *Biomedicines* 8 (5):118.

Al-Asmari AK, Islam M, and Al-Zahrani, AM. 2016. In vitro analysis of the anticancer properties of scorpion venom in colourectal and breast cancer cell lines. *Oncology Letters* 11 (2):1256–1262.

Aley KO, and Levine, JD. 2002. Different peripheral mechanisms mediate enhanced nociception in metabolic/toxic and traumatic painful peripheral neuropathies in the rat. *Neuroscience* 111 (2):389–397.

Attal N, Lanteri-Minet M, Laurent B, Fermanian J, and Bouhassira D. 2011. The specific disease burden of neuropathic pain: results of a French nationwide survey. *Pain* 152 (12):2836–2843.

Bach JF. 2002. The effect of infections on susceptibility to autoimmune and allergic diseases. *New England Journal of Medicine* 347 (12):911–920.

Baron R, Binder A, and Wasner G. 2010. Neuropathic pain: diagnosis, pathophysiological mechanisms, and treatment. *The Lancet Neurology* 9 (8):807–819.

Besson JM 1999. The neurobiology of pain. *The Lancet* 353 (9164).1610–1615.

Bouafir Y, Ait-Lounis A, and Laraba-Djebari F. 2017. Improvement of function and survival of pancreatic beta-cells in streptozotocin-induced diabetic model by the scorpion venom fraction F1. *Toxin Reviews* 36 (2):101–108.

Cao Z, Di Z, Wu Y, and Li W. 2014. Overview of scorpion species from China and their toxins. *Toxins* 6 (3):796–815.

Carbone E, Wanke E, Prestipino G, Possani LD, and Maelicke A. 1982. Selective blockage of voltage-dependent K^+ channels by a novel scorpion toxin. *Nature* 296 (5852):90–91.

Chandy KG, DeCoursey TE, Cahalan MD, McLaughlin C, and Gupta S. 1984. Voltage-gated potassium channels are required for human T lymphocyte activation. *The Journal of Experimental Medicine* 160 (2):369–385.

Chandy KG, Wulff H, Beeton C, Pennington M, Gutman GA, and Cahalan MD. 2004. K^+ channels as targets for specific immunomodulation. *Trends in Pharmacological Sciences* 25 (5):280–289.

Chatenoud L. 2016. Precision medicine for autoimmune disease. *Nature Biotechnology* 34 (9):930–932.

Chen Q, Yang P, Lin Q, Pei J, Jia Y, Zhong Z, and Wang S. 2021. Effects of scorpion venom heat-resistant peptide on the hippocampal neurons of kainic acid-induced epileptic rats. *Brazilian Journal of Medical Biological Research* 54 (5):e10717

Chen R, and Chung SH. 2012. Engineering a potent and specific blocker of voltage-gated potassium channel Kv1. 3, a target for autoimmune diseases. *Biochemistry* 51 (9):1976–1982.

Chen Y, Chen SR, Chen H, Zhang J,and Pan HL. 2019. Increased α2δ-1–NMDA receptor coupling potentiates glutamatergic input to spinal dorsal horn neurons in chemotherapy-induced neuropathic pain. *Journal of Neurochemistry* 148 (2):252–274.

Cologna CT, Marcussi S, Giglio JR, Soares AM, and Arantes EC. 2009. *Tityus serrulatus* scorpion venom and toxins: an overview. *Protein Peptide Letters* 16 (8):920–932.

Conde R, Zamudio FZ, Rodríguez MH, and Possani LD. 2000. Scorpine, an anti-malaria and anti-bacterial agent purified from scorpion venom. *FEBS Letters* 471 (2–3):165–168.

Crest M, Jacquet G, Gola M, Zerrouk H, Benslimane A, Rochat H, Mansuelle P, and Martin-Eauclaire MF. 1992. Kaliotoxin, a novel peptidyl inhibitor of neuronal BK-type Ca ($^{2+}$)-activated K^+ channels characterized from *Androctonus mauretanicus mauretanicus* venom. *Journal of Biological Chemistry* 267 (3):1640–1647.

Dai C, Ma Y, Zhao Z, Zhao R, Wang Q, Wu Y, Cao Z, and Li W. 2008. Mucroporin, the first cationic host defense peptide from the venom of *Lychas mucronatus*. *Antimicrobial Agents
Chemotherapy* 52 (11):3967–3972.

DeCoursey TE, George Chandy K, Gupta S, and Cahalan MD. 1984. Voltage-gated K⁺ channels
in human T lymphocytes: a role in mitogenesis? *Nature* 307 (5950):465–468.

de Oliveira Pimentel PM, de Assis DR, Gualdrón-Lopez M, Barroso A, Brant F, Leite
PG, de Lima Oliveira BC, Esper L, McKinnie SM, Vederas JC, do Nascimento
Cordeiro M. 2021. *Tityus serrulatus* scorpion venom as a potential drug source for
Chagas' disease: trypanocidal and immunomodulatory activity. *Clinical Immunology*
226:108713.

Desales-Salazar E, Khusro A, Cipriano-Salazar M, Barbabosa-Pliego A, and Rivas-Caceres
RR. 2020. Scorpion venoms and associated toxins as anticancer agents: update on their
application and mechanism of action. *Journal of Applied Toxicology* 40 (10):1310–1324.

DeSantis CE, Lin CC, Mariotto AB, Siegel RL, Stein KD, Kramer JL, Alteri R, Robbins AS,
and Jemal A. 2014. Cancer treatment and survivorship statistics, 2014. *CA: A Cancer
Journal for Clinicians* 64 (4):252–271.

Deshane J, Garner CC, and Sontheimer H. 2003. Chlorotoxin inhibits glioma cell invasion via
matrix metalloproteinase-2. *Journal of Biological Chemistry* 278 (6):4135–4144.

Díaz-García A, Morier-Díaz L, Frión-Herrera Y, Rodríguez-Sánchez H, Caballero-Lorenzo
Y, Mendoza-Llanes D, Riquenes-Garlobo Y, Fraga-Castro JA. 2013. In vitro anticancer
effect of venom from Cuban scorpion *Rhopalurus junceus* against a panel of human
cancer cell lines. *Journal of Venom Research* 4:5.

Domingos Possani L, Martin BM, and Svendsen IB. 1982. The primary structure of
noxiustoxin: A K⁺ channel blocking peptide, purified from the venom of the scorpion
Centruroides noxius Hoffmann. *Carlsberg Research Communications* 47:285–289.

Doth AH, Hansson PT, Jensen MP, and Taylor RS. 2010. The burden of neuropathic pain: a
systematic review and meta-analysis of health utilities. *Pain* 149 (2):338–344.

Ehret-Sabatier L, Loew D, Goyffon M, Fehlbaum P, Hoffmann JA, van Dorsselaer A, and
Bulet P. 1996. Characterization of novel cysteine-rich antimicrobial peptides from scorpion blood. *Journal of Biological Chemistry* 271 (47):29537–29544.

El-Ghlban S, Kasai T, Shigehiro T, Yin HX, Sekhar S, Ida M, Sanchez A, Mizutani A, Kudoh
T, Murakami H, and Seno M. 2014. Chlorotoxin-Fc fusion inhibits release of MMP-2
from pancreatic cancer cells. *BioMed Research International* 2014. DOI: https://doi.org/
10.1155/2014/152659

Fan Z, Cao L, He Y, Hu J, Di Z, Wu Y, Li W, and Cao Z. 2011. Ctriporin, a new anti-methicillin-
resistant *Staphylococcus aureus* peptide from the venom of the scorpion *Chaerilus
tricostatus*. *Antimicrobial Agents Chemotherapy* 55 (11):5220–5229.

García-Gómez BI, Coronas FIV, Restano-Cassulini R, Rodríguez RR, and Possani LD. 2011.
Biochemical and molecular characterization of the venom from the Cuban scorpion
Rhopalurus junceus. *Toxicon* 58 (1):18–27.

Ghosh R, Alajbegovic A, and Gomes AV. 2015. NSAIDs and cardiovascular diseases: role of
reactive oxygen species. *Oxidative Medicine Cellular Longevity* 2015:536962.

Gomez MV, Romano-Silva MA, and Prado MAM. 1995. Effects of tityus toxin on central nervous system. *Journal of Toxicology: Toxin Reviews* 14 (3):437–456.

Guida F, Luongo L, Aviello G, Palazzo E, De Chiaro M, Gatta L, Boccella S, Marabese I,
Zjawiony JK, Capasso R, and Izzo AA. 2012. Salvinorin A reduces mechanical
allodynia and spinal neuronal hyperexcitability induced by peripheral formalin injection. *Molecular Pain* 8:60.

Guo X, Ma C, Du Q, Wei R, Wang L, Zhou M, Chen T, and Shaw C. 2013. Two peptides,
TsAP-1 and TsAP-2, from the venom of the Brazilian yellow scorpion, *Tityus*

serrulatus: evaluation of their antimicrobial and anticancer activities. *Biochimie* 95 (9).1784–1794.

Gupta RA, Shah N, Wang KC, Kim J, Horlings HM, Wong DJ, Tsai MC, Hung T, Argani P, Rinn JL, and Wang Y. 2010. Long non-coding RNA HOTAIR reprograms chromatin state to promote cancer metastasis. *Nature* 464 (7291):1071–1076.

Gupta SD, Debnath A, Saha A, Giri B, Tripathi G, Vedasiromoni JR, Gomes A, and Gomes A. 2007. Indian black scorpion (*Heterometrus bengalensis Koch*) venom induced antiproliferative and apoptogenic activity against human leukemic cell lines U937 and K562. *Leukemia Research* 31 (6):817–825.

Gwee MCE, Cheah LS, and Gopalakrishnakone P. 1995. Involvement of the L-arginine-nitric oxide synthase pathway in the relaxant responses of the rat isolated anococcygeus muscle to a scorpion (*Leiurus quinquestriatus quinquestriatus*) venom. *Toxicon* 33 (9):1141–1150.

Harrison PL, Abdel-Rahman MA, Miller K, and Strong PN. 2014. Antimicrobial peptides from scorpion venoms. *Toxicon* 88:115–137.

Heinen, T-E, and da Veiga, ABG. 2011. Arthropod venoms and cancer. *Toxicon* 57 (4):497–511.

Hetru C, Letellier L, Oren Z, Hoffmann JA, and Shai Y. 2000. Androctonin, a hydrophilic disulphide-bridged non-haemolytic anti-microbial peptide: a plausible mode of action. *Biochemical Journal* 345 (3):653–664.

Hmed BN, Serria HT, and Mounir ZK. 2013. Scorpion peptides: potential use for new drug development. *Journal of Toxicology* 2013. DOI. https://doi.org/10.1155/2013/958797

Hu Y, Meng B, Yin S, Yang M, Li Y, Liu N, Li S, Liu Y, Sun D, Wang S, and Wang Y. 2022. Scorpion venom peptide HsTx2 suppressed PTZ-induced seizures in mice via the circ_0001293/miR-8114/TGF-β2 axis. *Journal of Neuroinflammation* 19 (1):284.

Ia A. 1999. The likely worldwide increase in erectile dysfunction between 1995 and 2025 and some possible policy consequences. *BJU International* 84:450–456.

Inoue K, and Tsuda M. 2018. Microglia in neuropathic pain: cellular and molecular mechanisms and therapeutic potential. *Nature Reviews Neuroscience* 19 (3):138–152.

Inyang KE, Szabo-Pardi T, Wentworth E, McDougal TA, Dussor G, Burton MD, Price TJ. 2019. The antidiabetic drug metformin prevents and reverses neuropathic pain and spinal cord microglial activation in male but not female mice. *Pharmacological Research* 139:1–16.

Jager S, Bucci C, Tanida I, Ueno T, Kominami E, Saftig P, and Eskelinen EL. 2004. Role for Rab7 in maturation of late autophagic vacuoles. *Journal of Cell Science* 117 (20):4837–4848.

Jonas P, Vogel W, Arantes EC, and Giglio JR. 1986. Toxin gamma of the scorpion *Tityus serrulatus* modifies both activation and inactivation of sodium permeability of nerve membrane. *Pflugers Archiv: European Journal of Physiology* 407 (1):92–99.

Joseph B, and George J. 2012. Scorpion toxins and its applications. *International Journal of Toxicological and Pharmacological Research* 4 (3):57–61.

Khosravi M, Mayahi M, Kaviani F, and Nemati M. 2017. The Effects of Isolated Fractions of *Mesobuthus eupeus* Scorpion venom on humoral immune response. *Journal of Arthropod-Borne Diseases* 11 (4):497.

Kiguchi N, Kobayashi Y, and Kishioka S. 2012. Chemokines and cytokines in neuroinflammation leading to neuropathic pain. *Current Opinion in Pharmacology* 12 (1):55–61.

Kirsch GE, Skattebøl A, Possani LD, and Brown AM. 1989. Modification of Na channel gating by an alpha scorpion toxin from *Tityus serrulatus*. *The Journal of General Physiology* 93 (1):67–83.

Kumar A, Kaur H, and Singh A. 2018. Neuropathic pain models caused by damage to central or peripheral nervous system. *Pharmacological Reports* 70 (2):206–216.

Lagarto A, Bueno V, Pérez MR, Rodríguez CC, Guevara I, Valdés O, Bellma A, Gabilondo T, and Padrón AS. 2020. Safety evaluation of the venom from scorpion *Rhopalurus junceus*: Assessment of oral short term, subchronic toxicity and teratogenic effect. *Toxicon* 176:59–66.

Langley PC, Van Litsenburg C, Cappelleri JC, and Carroll D. 2013. The burden associated with neuropathic pain in Western Europe. *Journal of Medical Economics* 16 (1):85–95.

Lima EG, and Freire-Maia L. 1977. Cardiovascular and respiratory effects induced by intracerebroventricular injection of scorpion toxin (tityustoxin) in the rat. *Toxicon* 15 (3):225–234.

Lima WG, Maia CQ, de Carvalho TS, Leite GO, Brito JC, Godói IP, de Lima ME, and Ferreira JM. 2022. Animal venoms as a source of antiviral peptides active against arboviruses: a systematic review. *Archives of Virology* 167 (9):1763–1772.

Loeser JD, and Treede RD. 2008. The Kyoto protocol of IASP basic pain terminology. *Pain* 137 (3):473–477.

Lu X, Lu D, Scully MF, and Kakkar VV. 2006. Integrins in drug targeting-RGD templates in toxins. *Current Pharmaceutical Design* 12 (22):2749–2769.

Massensini AR, Moraes-Santos T, Gomez MV, and Romano-Silva MA. 1998. Alpha-and beta-scorpion toxins evoke glutamate release from rat cortical synaptosomes with different effects on [Na$^+$] i and [Ca^{2+}] i. *Neuropharmacology* 37 (3):289–297.

McLane MA, Joerger T, and Mahmoud A. 2008. Disintegrins in health and disease. *Journal of Biological* 13 (1):6617–37.

Mirshafiey A. 2007. Venom therapy in multiple sclerosis. *Neuropharmacology* 53 (3):353–361.

Mishal R, Tahir HM, Zafar K, and Arshad M. 2013. Anti-cancerous applications of scorpion venom. *International Journal of Biological & Pharmaceutical Research* 4 (5):356–360.

Mohamed AA, Omar H, Ghaffar MF, Marie MS, Ramadan ME, Talima SM, Daly ME, and Mahmoud S. 2019. Single nucleotide polymorphism in adiponectin gene and risk of pancreatic adenocarcinoma. *Asian Pacific journal of cancer prevention: APJCP* 20 (1):139.

Moon JY, Song S, Yoon SY, Roh DH, Kang SY, Park JH, Beitz AJ, and Lee JH. 2012. The differential effect of intrathecal Nav1.8 blockers on the induction and maintenance of capsaicin- and peripheral ischemia-induced mechanical allodynia and thermal hyperalgesia. *Anesthesia and Analgesia* 114:215–223.

Niederberger E, and Geisslinger G. 2013. Proteomics and NF-κB: an update. *Expert Review of Proteomics* 10 (2):189–204.

Newman DJ, and Cragg GM. 2012. Natural products as sources of new drugs over the 30 years from 1981 to 2010. *Journal of Natural Products* 75 (3):311–335.

Nunes KP, Torres FS, Borges MH, Matavel A, Pimenta AMC, and De Lima ME. 2013. New insights on arthropod toxins that potentiate erectile function. *Toxicon* 69:152–159.

Nunes KP, and Webb RC. 2012. Mechanisms in erectile function and dysfunction: an overview. In *Erectile Dysfunction-Disease-Associated Mechanisms Novel Insights into Therapy*, edited by KP Nunes. London, UK: Intech Open. ISBN 978-953-51-0199-4.

Paksoy N, and Tetik BK. 2022. Spondyloarthritis recovering after scorpion sting: a case report. *Medicine* 11 (2):883–885.

Peigneur S, Cologna CT, Cremonez CM, Mille BG, Pucca MB, Cuypers E, Arantes EC, and Tytgat J. 2015. A gamut of undiscovered electrophysiological effects produced by *Tityus serrulatus* toxin 1 on NaV-type isoforms. *Neuropharmacology* 95:269–277.

Pennington MW, Czerwinski A, and Norton RS. 2018. Peptide therapeutics from venom: current status and potential. *Bioorganic Medicinal Chemistry* 26 (10):2738–2758.

Perez-Verdaguer M, Capera J, Serrano-Novillo C, Estadella I, Sastre D, and Felipe A. 2016. The voltage-gated potassium channel Kv1. 3 is a promising multitherapeutic target against human pathologies. *Expert Opinion on Therapeutic Targets* 20 (5):577–591.

Pucca MB, Cerni FA, Peigneur S, Bordon KCF, Tytgat J, and Arantes EC. 2015. Revealing the function and the structural model of Ts4: insights into the "non-toxic" toxin from *Tityus serrulatus* venom. *Toxins* 7 (7):2534–2550.

Rao VR, Perez-Neut M, Kaja S, and Gentile S. 2015. Voltage-gated ion channels in cancer cell proliferation. *Cancers* 7 (2):849–875.

Rapôso C. 2017. Scorpion and spider venoms in cancer treatment: state of the art, challenges, and perspectives. *Journal of Clinical Translational Research* 3 (2):233.

Raza M, Shaheen F, Choudhary MI, Sombati S, Rafiq A, Suria A, and DeLorenzo RJ. 2001. Anticonvulsant activities of ethanolic extract and aqueous fraction isolated from *Delphinium denudatum*. *Journal of Ethnopharmacology* 78 (1):73–78.

Rigo FK, Bochi GV, Pereira AL, Adamante G, Ferro PR, De Prá SD, Milioli AM, Damiani AP, da Silveira Prestes G, Dalenogare DP, Chávez-Olórtegui C. 2019. TsNTxP, a non-toxic protein from *Tityus serrulatus* scorpion venom, induces antinociceptive effects by suppressing glutamate release in mice. *European Journal of Pharmacology* 855:65–74.

Rose NR. 2016. Prediction and prevention of autoimmune disease in the 21st century: a review and preview. *American Journal of Epidemiology* 183 (5):403–406.

Ru Q, Tian X, Wu YX, Wu RH, Pi MS, and Li CY. 2014. Voltage-gated and ATP-sensitive K$^+$ channels are associated with cell proliferation and tumorigenesis of human glioma. *Oncology Reports* 31 (2):842–848.

Salem ML, Shoukry NM, Teleb WK, Abdel-Daim MM, and Abdel-Rahman MA. 2016. In vitro and in vivo antitumor effects of the Egyptian scorpion *Androctonus amoreuxi* venom in an Ehrlich ascites tumor model. *Springerplus* 5:1–12.

Shen B, Cao Z, Li W, Sabatier JM, and Wu Y. 2017. Treating autoimmune disorders with venom-derived peptides. *Expert Opinion on Biological Therapy* 17 (9):1065–1075.

Song YM, Tang XX, Chen XG, Gao BB, Gao E, Bai L, and Lv XR. 2005. Effects of scorpion venom bioactive polypolypeptides on platelet aggregation and thrombosis and plasma 6-keto-PG F1α and TXB2 in rabbits and rats. *Toxicon* 46 (2):230–235.

Srinivasan KN, Nirthanan S, Sasaki T, Sato K, Cheng B, Gwee MC, Kini RM, and Gopalakrishnakone P. 2001. Functional site of bukatoxin, an α-type sodium channel neurotoxin from the Chinese scorpion (*Buthus martensi Karsch*) venom: probable role of the 52PDKVP56 loop. *FEBS letters* 494 (3):145–149.

Tanner MR, Pennington MW, Chamberlain BH, Huq R, Gehrmann EJ, Laragione T, Gulko PS, and Beeton C. 2018. Targeting KCa1. 1 channels with a scorpion venom peptide for the therapy of rat models of rheumatoid arthritis. *Journal of Pharmacology Experimental Therapeutics* 365 (2):227–236.

Teixeira CE, Teixeira SA, Antunes E, and De Nucci G. 2001. The role of nitric oxide on the relaxations of rabbit corpus cavernosum induced by *Androctonus australis* and *Buthotus judaicus* scorpion venoms. *Toxicon* 39 (5):633–639.

Toda N, Ayajiki K, Okamura T. 2005. Nitric oxide and penile erectile function. *Pharmacology Therapeutics* 106 (2):233–266.

Turner KL, and Sontheimer H. 2014. Cl$^-$ and K$^+$ channels and their role in primary brain tumour biology. *Philosophical Transactions of the Royal Society B: Biological Sciences* 369 (1638):20130095.

Villalonga N, Ferrere JC, Argiles JM, Condom E, and Felipe A. 2007. Potassium channels are a new target field in anticancer drug design. *Recent Patents on Anti-Cancer Drug Discovery* 2 (3):212–223.

Wang J, and Xiang M. 2013. Targeting potassium channels K v1. 3 and KC a3. 1: routes to selective immunomodulators in autoimmune disorder treatment? *Pharmacotherapy: The Journal of Human Pharmacology Drug Therapy* 33 (5):515–528.

Wang JD, Narui T, Takatsuki S, Hashimoto T, Kobayashi F, Ekimoto H, Abuki H, Niijima K, and Okuyama T. 1991. Hematological studies on naturally occurring substances. VI. Effects of an animal crude drug "chan su" (bufonis venenum) on blood coagulation, platelet aggregation, fibrinolysis system and cytotoxicity. *Chemical Pharmaceutical Bulletin* 39 (8):2135–2137.

Wang X, Zhang S, Zhu Y, Zhang Z, Sun M, Cheng J, Xiao Q, Li G, and Tao J. 2020. Scorpion toxins from *Buthus martensii Karsch* (BmK) as potential therapeutic agents for neurological disorders: state of the art and beyond. In Medical Toxicology, edited by P Erkekoglu, and T Ogawa. London, UK: Intech Open. ISBN 978-1-83880-278-3.

Xie J, and Herbert TP. 2012. The role of mammalian target of rapamycin (mTOR) in the regulation of pancreatic β-cell mass: implications in the development of type-2 diabetes. *Cellular Molecular Life Sciences* 69:1289–1304.

Xie JD, Chen SR, and Pan HL. 2017. Presynaptic mGluR5 receptor controls glutamatergic input through protein kinase C–NMDA receptors in paclitaxel-induced neuropathic pain. *Journal of Biological Chemistry* 292 (50):20644–20654.

Yu QT, and Meng ZB. 2016. Treatment of advanced breast cancer with a combination of highly agglutinative staphylococcin and vinorelbine-based chemotherapy. *European Review for Medical and Pharmacological Sciences* 20 (16):3465–3468.

Zargan J, Sajad M, Umar S, Naime M, Ali S, and Khan HA. 2011. Scorpion (*Androctonus crassicauda*) venom limits growth of transformed cells (SH-SY5Y and MCF-7) by cytotoxicity and cell cycle arrest. *Experimental Molecular Pathology* 91 (1):447–454.

Zeng XC, Wang SX, Zhu Y, Zhu SY, and Li WX. 2004. Identification and functional characterization of novel scorpion venom peptides with no disulfide bridge from *Buthus martensii Karsch*. *Peptides* 25 (2):143–150.

Zhang YY, Wu LC, Wang ZP, Wang ZX, Jia Q, Jiang GS, and Zhang WD. 2009. Antiproliferation effect of polypeptide extracted from scorpion venom on human prostate cancer cells in vitro. *Journal of Clinical Medicine Research* 1 (1):24.

Zhao Y, Huang J, Yuan X, Peng B, Liu W, Han S, and He X. 2015a. Toxins targeting the Kv1. 3 channel: potential immunomodulators for autoimmune diseases. *Toxins* 7 (5):1749–1764.

Index

For Product Safety Concerns and Information please contact our EU
representative GPSR@taylorandfrancis.com
Taylor & Francis Verlag GmbH, Kaufingerstraße 24, 80331 München, Germany

www.ingramcontent.com/pod-product-compliance
Lightning Source LLC
Chambersburg PA
CBHW060249230326
41458CB00094B/1586